Introduction

The Brandenburg D.I was the first biplane fighter of the LFT, or *Luftfahrtruppe,* the Austro-Hungarian air service. The technical development of this interesting aircraft has been detailed in *Austro-Hungarian Army Aircraft of World War One* by Peter M. Grosz, George Haddow, and Peter Scheimer. The development text that follows has been excerpted from that book by kind permission of Peter M. Grosz, who has also generously contributed many additional photographs from his extensive archives.

The detailed operational history of the Brandenburg D.I is the work of Karl Meindl and Walter Schroeder, who contributed most of the photographs for that section together with the color references. This book is therefore the most complete history of this significant aircraft ever presented.

The Brandenburg D.I, frequently called the KD for *Kampfdoppeldecker,* or biplane fighter, was a transitional fighter. It offered its pilots better performance than the Fokker monoplanes that had been the early mainstay of the LFT, and lead to the more capable Aviatik D.I, Albatros D.II and D.III, and Phönix D.I, D.II, and D.IIa biplane fighters that followed it into service. The KD was sturdy and streamlined, its unique 'star' strut arrangement eliminating the drag-producing bracing wires used by most biplanes. This made it fast for its time, a critical quality in a fighter.

Unfortunately, the 'star' struts were also heavy. Their weight, coupled with the weight of the water-cooled engine, gave it a poor rate of climb. The over-wing armament used by most KD fighters was troublesome and unreliable, a trait it shared with the similarly-armed Nieuport 11 and 17 fighters that were its frequent adversaries. Gun jams were numerous and clearing them impossible. The KD's greatest weakness was its poor maneuverability and handling qualities, which were dangerous to its pilots and put it at a distinct disadvantage in the close-in, high-G maneuvering combats or 'dogfights' that characterized the first air war. However, the KD was not alone in having dangerous handling qualities; even the highly-regarded Sopwith Camel had such treacherous handling qualities that more of its pilots were killed in training and operational accidents than in air combat. The Camel's redeeming feature, for those pilots who survived flying it long enough to master its idiosyncracies, was its excellent if neurotic maneuverability, a quality the KD unfortunately lacked.

The Brandenburg D.I was developed into the excellent Phönix D.I fighter, which eliminated the heavy strut arrangement and combined a more powerful engine with twin synchronized machine guns to give it the improved climb, maneuverability, handling qualities, and firepower required of a front-line fighter by the last year of the war.

Above: A smiling Karl Kaszala stands before KD 28.11 after a bad landing at *Flik* 41/J. Kaszala had 8 confirmed victories and scored his fifth victory in this aircraft on 20 May 1917. Photo: Karl Meindl.

Left: KD 28.06 flown by 32-victory ace Julius Arigi, *Fluggeschwader* I, 1917, Isonzo front, spirng 1917. The lower left wingtip is in the shadow of the photo aircraft. Photo: J. Bangerl via Karl Meindl.

Below: Godwin Brumowski's dark KD at left, probably 28.69, with Linke-Crawford's Albatros D.III (Oef) 153.11 at right on Ajello airfield, *Flik* 41/J. Photo: James Kerr.

Development

Oberst Uzelac, commander of the LFT, informed his staff in August 1916 that "the Fokker fighter is outmoded; therefore I have authorized the production of a light biplane fighter that is equal to all combat requirements." He was referring to the Brandenburg D.I (Type KD), the prototype of which (60.55) was currently undergoing evaluation by *Flik* 26. Production began in May 1916, and by the time the formal contracts were signed assembly at Briest and production preparation at Phönix were well advanced. A total of 122 D.I fighters were ordered:

Qty	Manufacturer	Series No.	Engine	Order Date
24	Phönix	28.01–24	Dm 185	24 June 1916
20	Brandenburg	65.50–69	Dm 160	25 August 1916
30	Brandenburg	65.70–99	Dm 150	25 August 1916
24	Phönix	28.25–48	Dm 185	3 December 1916
24	Phönix	28.49–72	Dm 185	29 January 1917

The D.I (or KD as it was commonly called) is easily identified by the unorthodox pyramid struts, originally proposed by Professor Knoller as a means of attaining great strength and lowering drag by eliminating wire bracing. In static load tests the pyramid cellule attained a safety factor of 9, but at the expense of a structure heavier than that of conventional design. Two Brandenburg-built D.I series were produced, the series 65.50 to 65.69, powered by a 160 hp Daimler engine, and series 65.70 to 65.99 powered by a 150 hp Daimler engine. The D.I series 65.7 was lightened throughout but, given the reduced horsepower, demonstrated only marginal improvement in maneuverability and climb. Squadron introduction, slated for September 1916, was delayed by re-work after flight trials had revealed weak engine supports and excessive wing-tip vibration. One D.I series 65.5 fighter was accepted in October and the remaining 19 in December 1916. Series 65.7 acceptances began in November 1916 and the Phönix-built series 28 in January 1917.

A total of 72 Brandenburg D.I(Ph) fighters, numbered 28.01 to 28.72, were ordered from Phönix between June 1916 and January 1917. Using aircraft 65.57 as a sample, Phönix engineers made a number of significant changes. Power was increased by installing a 185 hp Daimler engine, which imparted a slight performance increase. The center-section structure was simplified and the pilot's seat was raised to enable gun aiming without leaning out of the cockpit. Acceptances, delayed by a propeller shortage, began in January 1917, and the first series 28 fighters arrived at *Fliks* 41/J and 42/J in February–March 1917. Being lighter and more powerful than their German counterpart, series 28 fighters possessed take-off characteristics that "did not engender fear that speed had to be maintained." The series 28 D.I(Ph) was vastly preferred over the series 65.5 and in the hands of pilots such as Brumowski, Kiss, or Linke-Crawford, scored the lion's share of the victories for this type. Yet for the average or novice fighter pilot, the Brandenburg D.I's tricky flight habits and propensity to spin at all altitudes made it an extremely hazardous aircraft.

Beginning with 28.51, Phönix-built aircraft were delivered with a lightened airframe. The large, rounded tail fin and rudder was either retro-fitted or installed on all production aircraft. Even so, when the last fighters were accepted in mid-1917, the Brandenburg D.I was already out-classed.

Above: The first Brandenburg KD prototype powered by a 160 hp Mercedes.

Above: The second KD prototype, 60.55 (ex 05.09) shows drag-reducing refinements such as an airfoil radiator, a raised rear fuselage, a refined engine cowling, and tapered strut ends.

Left: The unarmed Brandenburg D.I(Ph) 28.01 fighter during flight testing at Aspern in January 1917. The center section has been simplified and the strut attachement points are covered with cuffs. On 1 March 1917, *Hauptmann* Franz Rabitsch, commanding officer of *Fluggeschwader* I, lost his life when this aircraft crashed during a familiarization flight.

Writing in 1942, *Oberst* Adolf Heyrowsky, commander of *Flik* 19 (first to receive the new fighter) regarded the D.I obsolete on arrival. He felt that the 3000 meter ceiling was unacceptable and the tricky flight characteristics much too difficult for inexperienced pilots, especially since the D.I exhibited a sudden tendency to spin "with consequent casualties." At *Flik* 19, Heyrowsky grounded the D.I pending further instruction. Uzelac arrived unannounced on 7 November 1916 to personally discuss the problem, but Heyrowsky was absent from the airfield. Upon return, Heyrowsky faced a wrecked D.I (see page 53). Asking who had flown the machine against his strict orders, the hesitant answer was "*Oberst* Uzelac; he crashed violently and is in the local hospital." Fearing the worst, Heyrowsky rushed to the hospital to learn that Uzelac had received a severe concussion, but upon regaining consciousness had summoned his chauffeur and was gone.

So began the D.I's service history, the first fighter to be built in quantity for the LFT. In late 1916, the D.I series 65.5 and 65.7 were supplied to the Isonzo Front, namely *Fliks* 4, 12, 16, 19, 23, 28, 34, and 41/J and *Fluggeschwader* I, serving primarily as an escort fighter. The initial combat experience was universally discouraging. In February 1917, a *Flars* report stated that the D.I was unsatisfactory because of its poor climb and dangerous flight characteristics. "Until a short while ago, the KD was praised as being everything a pilot could wish for; today it carries the unflattering name of '*Totschläger*' (killer)." Assigned the task of making improvements, Dr. von Mises of *Flars* solicited the opinions of front-line pilots, one of whom was Julius Arigi, who had already installed a "homemade" fin and rudder on his machine, not unlike the final version that was tested on 3 April 1917 and adopted. Arigi was awarded a substantial cash prize. Replacement kits were dispatched to the front on 17 May 1917. At best it was a stopgap measure.

The prevailing attitude of operational units is best illustrated in an excerpt from a *Flik* 23 report (20 May 1917) recommending the fighter's withdrawal. The incident described was not uncommon:

> *Leutnant* Mirko Vrbanic, one of our best KD pilots, stalled during an escort mission at 3000 meters and was able to regain control just short of crashing. In the pilots' unanimous opinion, they cannot give full attention to the combat at hand if they are so totally occupied with controlling the aircraft. In addition, the climb is so slow that the fighter must take off well before our observation aircraft in order to reach escort altitude. The KD's ceiling is much inferior to that of enemy Nieuports, which generally operate between 4 and 5000 meters.

Even crack fighter units such as *Flik* 41/J reported that:

> ...the KD is absolutely useless (June 1917)...the best pilots (and only they can fly the type) are shackled, ruin their nerves, and perish in crashes over the airfield, without their expert skill achieving anything.

The non-combat losses far exceeded those in combat: *Flik* 41/J lost eleven D.I fighters to accidents alone in July. That month, *Flik* 42/J reported that "The KD is inferior to enemy fighters. The enemy, aware that our fighters lose altitude in a turn, utilizes this to his advantage." One wonders how the novice fighter pilot must have felt hearing the D.I popularly referred to as *der fliegender Sarg* (the flying coffin).

The KD was usually armed with the heavy (37 kg–82 lb) Type II VK gun canister (aptly nicknamed "the baby coffin") on the top wing. This bulky installation "further reduced the existing marginal flight characteristics." Several interim solutions of storing the ammunition in the fuselage and fitting an oblique-firing gun were tested but not used in combat. Later a few Phönix-built KDs were armed with one synchronized gun installed below the exhaust manifold.

Damned by most, the D.I achieved marginal success in the hands of expert pilots, among them Arigi, Brumowski, Fiala, Kiss, and Linke-Crawford. But almost all victories were gained in the Phönix-built version. An analysis of 68 confirmed D.I victories shows that only three can be attributed to pilots flying the Brandenburg-built fighter. Why this should be is not known, but perhaps the Phönix machine was of superior construction. In all fairness it must be said that the D.I was not entirely to blame for its deadly reputation. Fighter pilot training was rudimentary, undertaken in obsolete two seaters, with only a few students

Above: Another view of the first prototype KD.

Above, below, and below left: Additional views of 60.55 at *Flik* 26.

Above: 60.55 with the lower inner rear strut missing. Previously designated 05.09.
Below: Another view of 60.55, which was unarmed at this time.

fortunate enough to obtain advanced instruction in high-performance fighter-trainers.

On 26 May 1917, Uzelac, writing to the 5th Army, acknowledged the D.I's defects, but until the expected arrival

Above: Brandenburg D.I 65.60.

of new fighters in July it was impossible to make a change. The hard-pressed *Fliks* had little choice but to continue flying the D.I fighter. As of 1 July 1917, 41 Brandenburg and 49 Phönix-built D.I fighters were at the Front. As a last resort some *Fliks* modified the D.I for photo-reconnaissance work, achieving good results in a role where speed, not maneuverability, counted. As new fighters arrived, the Brandenburg D.I was gradually withdrawn from operational service, finding employment as *Flik*-based trainers for newly-arrived pilots or assigned to training units, notably *Fleks* 4, 5, and 6. In the fall of 1917, the remaining fighters were gradually withdrawn and assigned to training units in which they served until late 1918. The LFT ordered all Brandenburg D.I fighters removed from the front on 27 January 1918.

Brandenburg D.I Series 65.5 Specifications

Engine:	160 hp Daimler	
Wing:	Span Upper	8.50 m (27.89 ft)
	Span Lower	8.30 m (27.23 ft)
	Chord Upper	1.50 m (4.92 ft)
	Chord Lower	1.50 m (4.92 ft)
	Sweepback Upper	0 deg
	Sweepback Lower	0 deg
	Gap	1.60 m (5.25 ft)
	Stagger	0.50 m (1.64 ft)
	Total Wing Area	24.0 sq m (258 sq ft)
General:	Length	6.63 m (21.75 ft)
	Height	2.75 m (9.02 ft)
	Track	1.72 m (5.64 ft)
	Empty Weight	745 kg (1643 lb)
	Loaded Weight	1020 kg (2249 lb)
Maximum Speed:		185 km/hr (115 mph)
Climb:	1000m (3,281 ft) in	4 min 8 sec

Brandenburg D.I Series 65.7 Specifications

Engine:	150 hp Daimler	
Wing:	Span Upper	8.50 m (27.89 ft)
	Span Lower	8.30 m (27.23 ft)
	Chord Upper	1.50 m (4.92 ft)
	Chord Lower	1.50 m (4.92 ft)
	Dihedral Upper	0 deg
	Dihedral Lower	0.08 m (0.26 ft) at tip
	Sweepback Upper	0 deg
	Sweepback Lower	0 deg
	Gap	1.60 m (5.25 ft)
	Stagger	0.50 m (1.64 ft)
	Total Wing Area	25.21 sq m (271 sq ft)
General:	Length	6.63 m (21.75 ft)
	Height	2.80 m (9.19 ft)
	Track	1.72 m (5.64 ft)
	Empty Weight	672 kg (1482 lb)
	Loaded Weight	940 kg (2073 lb)
Maximum Speed:		179 km/hr (111 mph)
Climb:	1000m (3,281 ft) in	3 min 30 sec

Brandenburg D.I(Ph) Series 28 Specifications

Engine:	185 hp Daimler	
Wing:	Span Upper	8.50 m (27.89 ft)
	Span Lower	8.50 m (27.89 ft)
	Chord Upper	1.50 m (4.92 ft)
	Chord Lower	1.50 m (4.92 ft)
	Dihedral Upper	0 deg
	Dihedral Lower	1.8 deg
	Sweepback Upper	0 deg
	Sweepback Lower	0 deg
	Gap	1.60 m (5.25 ft)
	Stagger	0.50 m (1.64 ft)
	Total Wing Area	24.0 sq m (258 sq ft)
General:	Length	6.30 m (20.67 ft)
	Height	2.79 m (9.15 ft)
	Track	1.72 m (5.64 ft)
	Empty Weight	714 kg (1574 lb)
	Loaded Weight	1047 kg (2309 lb)
Maximum Speed:		175–185 km/hr (109–115 mph)
Climb:	1000m (3,281 ft) in	3 min 55 sec

Brandenburg D.I(Ph) Series 28.5 Specifications

Engine:	185 hp Daimler	
Wing:	Span Upper	8.50 m (27.89 ft)
	Span Lower	8.50 m (27.89 ft)
	Chord Upper	1.60 m (5.25 ft)
	Chord Lower	1.60 m (5.25 ft)
	Gap	1.60 m (5.25 ft)
	Stagger	0.50 m (1.64 ft)
	Total Wing Area	25.5 sq m (274 sq ft)
General:	Length	6.30 m (20.67 ft)
	Height	2.79 m (9.15 ft)
	Empty Weight	690 kg (1521 lb)
	Loaded Weight	958 kg (2112 lb)
Maximum Speed:		175–177 km/hr (109–110 mph)
Climb:	1000m (3,281 ft) in	4 min 30 sec

Brandenburg D.I Series 28

Designer: Ernst Heinkel of the Hansa-Brandenburg Flugzeugwerke.

Manufacturer: Österreichische Albatros Flugzeugwerke, from 1.1.1917 on, known as the Phönix Flugzeugwerke.

Years of Construction: 1916–1917.

Intended Purpose: Single-seater fighter, from February 1918 on, used only as a pursuit trainer.

Construction Method: Generally like Series 65.5; airfoil radiator by the Hirschfelder Fabrik (installed in 20 KDs), Norddeutsche Kühlerfabrik, as well as the Flugzeug- und Automobilbestandteil Fabrik. According to front-line reports, one type of radiator had failed to meet the requirements (but which one?). With Series 28, there were a conspicuous number of carburetor fires (engine fires) with heavy losses of personnel.

Engine: 185 hp Austro-Daimler 6-cylinder in-line engine.

Propeller: *Sigma* or German *Rahtjen* propeller.

Armament: One fixed, unsynchronized Schwarzlose M7/12 or M.16 machine gun in a VK II canister (also called Kann canisters after the designer, *Hptm.i.d.Res.* Dr. Leopold Kann). (For instance, 28.28 and 28.62 each had one M7/12 machine gun in this arrangement.) However, 28.06 and 28.08 of *FlG* I had two machine guns (one M7/12 and M.16 each) in a VK II canister. 28.26 (which was also used as a photo single-seater) and 28.30 of Flik 16 as well as 28.46 of *FlG* I had a tiltable Schwarzlose M7/12 machine gun on the upper wing, on an experimental basis. 28.37 of *Flik* 24 as well as 28.29 (presumably also 28.47) of *Fliegerdetachment Hauptmann Nikitsch* had a synchronized machine gun (Bernatzik synchronization). Five KDs of Series 28 were reputed to have synchronized machine guns.

Development: Licensed production of the German Hansa-Brandenburg KD by the *Öst. Albatros Flugzeugwerke* (renamed the *Phönix Flugzeugwerke* after 1.1.17) with the more powerful 185 hp Austro-Daimler 6-cylinder in-line engine. On the basis of experiences with the airplane prototype 28.47, all Series 28 aircraft from 28.49 on were given lightened wing cellules, the so-called "Nieuport" (after the model of captured Italian and French fighters) or "Sparmann" wing cellules (after the test pilot and designer *Lt.i.d.Res.* Edmund Sparmann of the Phönix Flugzeugwerke) as well as longer upper wings, shorter bottom wings, and modified ailerons, which were supposed to improve the aircraft's rate of climb performance. Furthermore, (differing from the German model), a raised fuselage and an enlarged rudder to improve the lateral stability and turn characteristics. In a turn, the KD types had to be flown not with the rudder, but with the ailerons.

Of the first 48 KD aircraft of Series 28, aircraft from 28.47 on were delivered with raised fuselages and expanded and raised rudders. Existing aircraft of this type were retrofitted. (This, by the way, was an invention of *Feldpilot* Arigi, for which he received a reward of 500 *Kronen* from *Oberst* Uzelac.) The retrofitting took place starting in late May of 1917. (Diary of *Feldpilot* Linke-Crawford.)

Orders: *Abteilung* 5/L of the Imperial Austro-Hungarian Ministry of War ordered 24 KDs of Type 28 with each of three orders (5/L 703/1916, 5/L 4328/1916, and 5/L 426/1917), of which the last series of 24 had the "Nieuport" wing cellule. The Type 28 aircraft were delivered from roughly the end of 1916 to July 1917. Due to delivery problems with the propellers, they first reached the front in February 1917. (Thus 28.06 was accepted on 2.2. and 28.46 on 28.8.1917. Those are the dates of their front-line delivery, and not of their hinterland acceptance!)

As a result of the delivery of Type 28 being delayed, the *k.u.k. 5. Armee* (Isonzo front) was allocated six Series 65.7 KDs which were actually only considered pursuit trainers on account of their 150 hp engines. In various albums, there are many photos of this type with the well-known machine-gun canister and, of course, "Kraxen" pictures, as well. However, with the existing lists of offensive patrols, deployments using this type cannot be proven.

Price: Converting the asking price into present-day currency is very problematic, which is why no attempt has been made to do so. Only one thing is certain: that the amounts asked were paid. The first 24 KD aircraft of Series 28 cost 28,000 *Kronen*; for the second series of 24 aircraft, 30,000 *Kronen* had to be paid, and for the last 24 machines, the price rose to 33,000 *Kronen*. The price was understood as being for *Drachen* ("kites" or aircraft without engines); neither the armament nor the machine-gun canisters were included in the sums quoted. Engines were considerably more expensive in Austro-Hungary than in Germany, but in return, the *Drachen* were somewhat cheaper.

Utilization: As single-seater fighters and escort aircraft by *Fliks* 2, 12, 23, 28, 32, 34, 35, 41/J, 42/J, 46/F, and *Fluggeschwader* I (later 101/G) on the Isonzo front; by *Fliks* 17, 21, 24, and 101/G on the south Tyrolean front; and by *Flik* 16 as escort fighters and photo single-seaters (Carinthian front). Also used by *Flik* 36 and *Fliegerdetachment Hauptmann Nikitsch* on the Romanian front.

At the end of February 1918, all existing KD aircraft of Series 28 were withdrawn from the front, and those airplanes still airworthy were then used only for training fighter pilots.

Despite its obvious shortcomings (poor take-off and landing characteristics, a miserable rate of climb, a lack of stability in air combat, innumerable engine defects, carburetor fires in particular and horrible armament), numerically this type was the prevailing type of Austro-Hungarian fighter up until the summer of 1917.

The successes described give rise to astonishment and amazement and can only be explained by the fact that absolutely first-rate pilots were sitting in the cockpits.

Technical Data of the First and Second Series of 24 Aircraft Each, Type 28: Wingspan, upper and lower, 8.5 and 8.3 m respectively; wing chord 1.5 m; length 6.35 m; height 2.79 m; wing area 24 sq.m; elevator 0.96 sq.m; rudder 0.45 sq.m; aileron 1.77 sq.m; track 1.72 m; propeller [diameter] 2.66 m; empty weight 714 kg; payload 130 kg; gasoline 93 kg; oil 24 kg; coolant 21 kg; loaded weight 1,047 kg; wing load 39.5 kg per sq.m.

Flight performance of 28.06 during its acceptance flight on 27 January 1917: v/max. 187 km/h; climbing time to 1,000 m, 3 minutes 55 seconds; flight duration 2.5 hours; take-off and landing run supposedly only 80 and 50 m respectively. The performance data given were achieved in the hinterland, under ideal conditions by first-rate test pilots. Reports from front-line airmen speak of a horizontal speed of approximately 150–160 km/h, so greatly was its performance reduced by installation of the machine-gun canisters.

Technical Data of the Third Series of 24 Aircraft of Type 28: (Only variating measurements and weights are given here.) Wing span, upper and lower, 8.6 m and 6 m respectively; length 6.3 m; height 2.72 m; wing area 17.5 sq.m; aileron 2 sq.m; empty weight 656 kg; gasoline 95 kg; oil 10 kg; coolant 20 kg; loaded weight 920 kg; wing load 47 kg per sq.m.

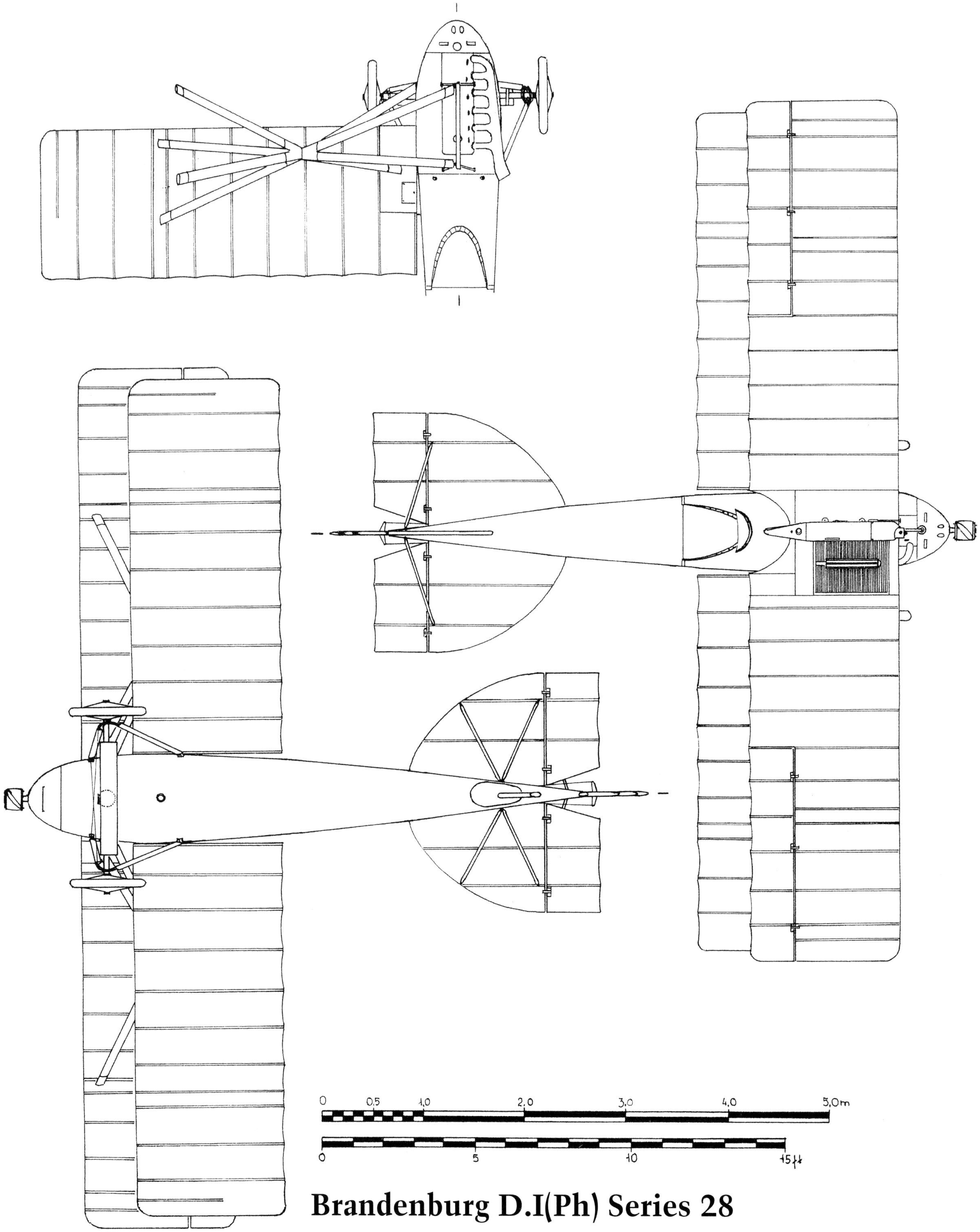

Brandenburg D.I(Ph) Series 28

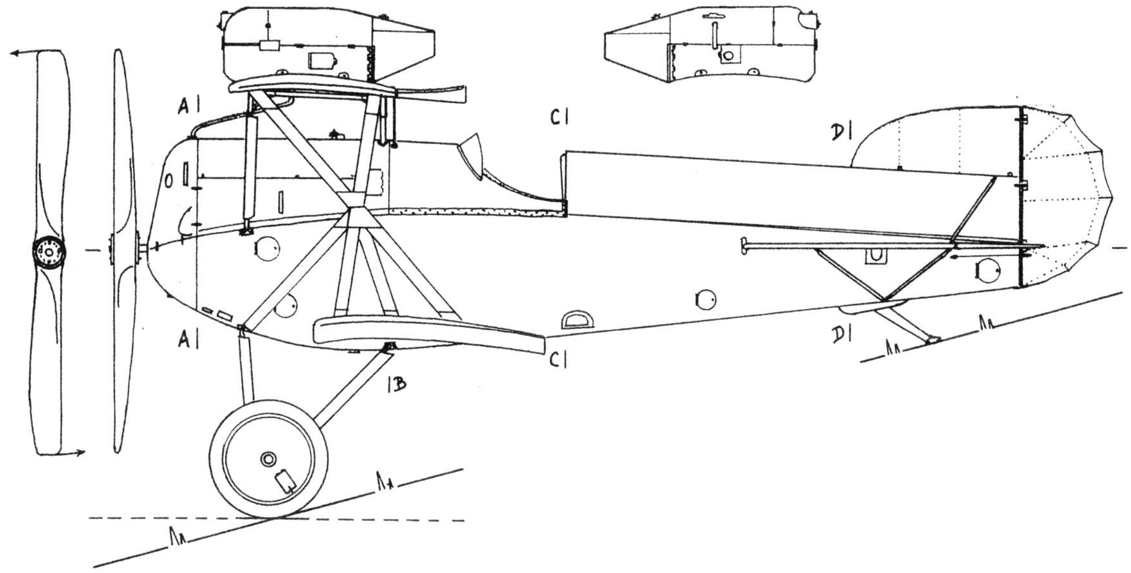

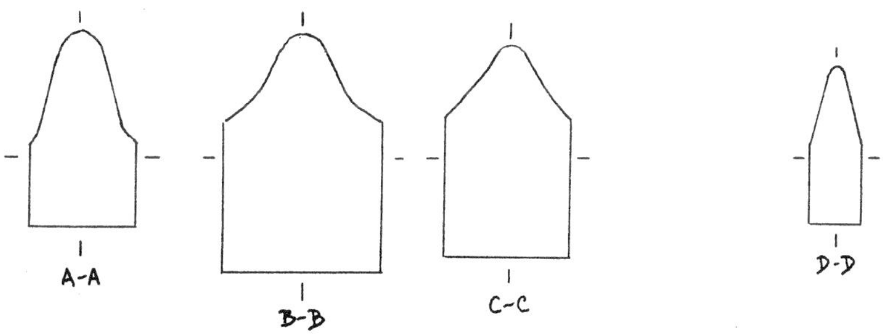

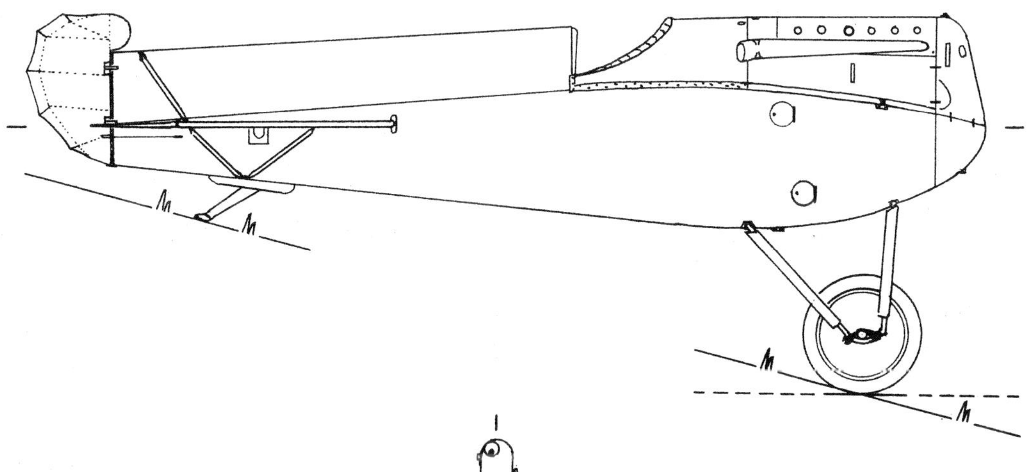

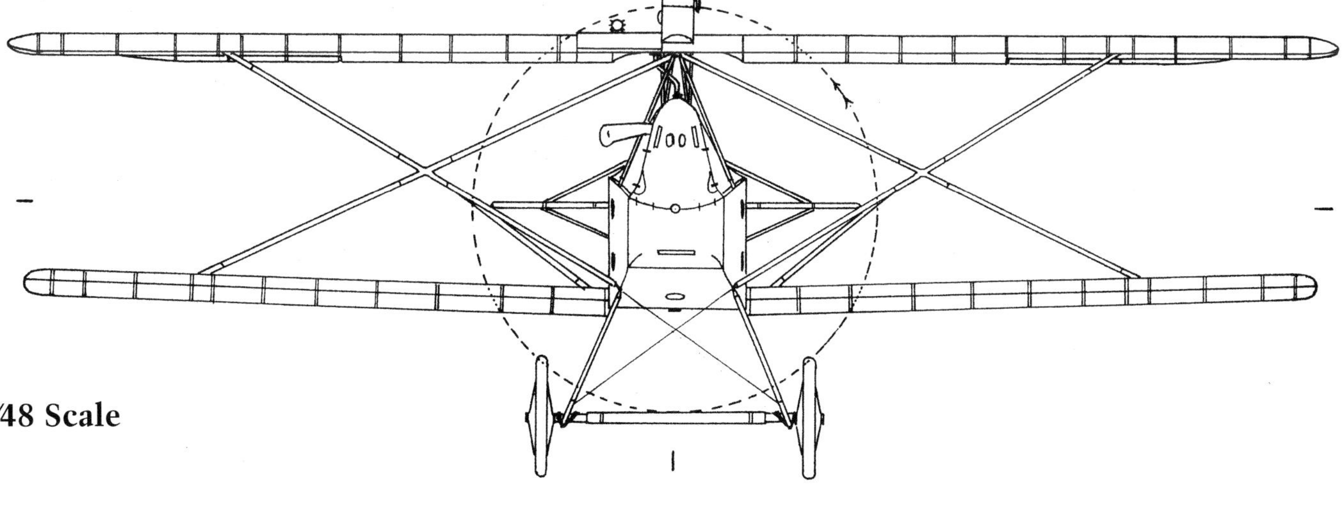

1⁄48 Scale

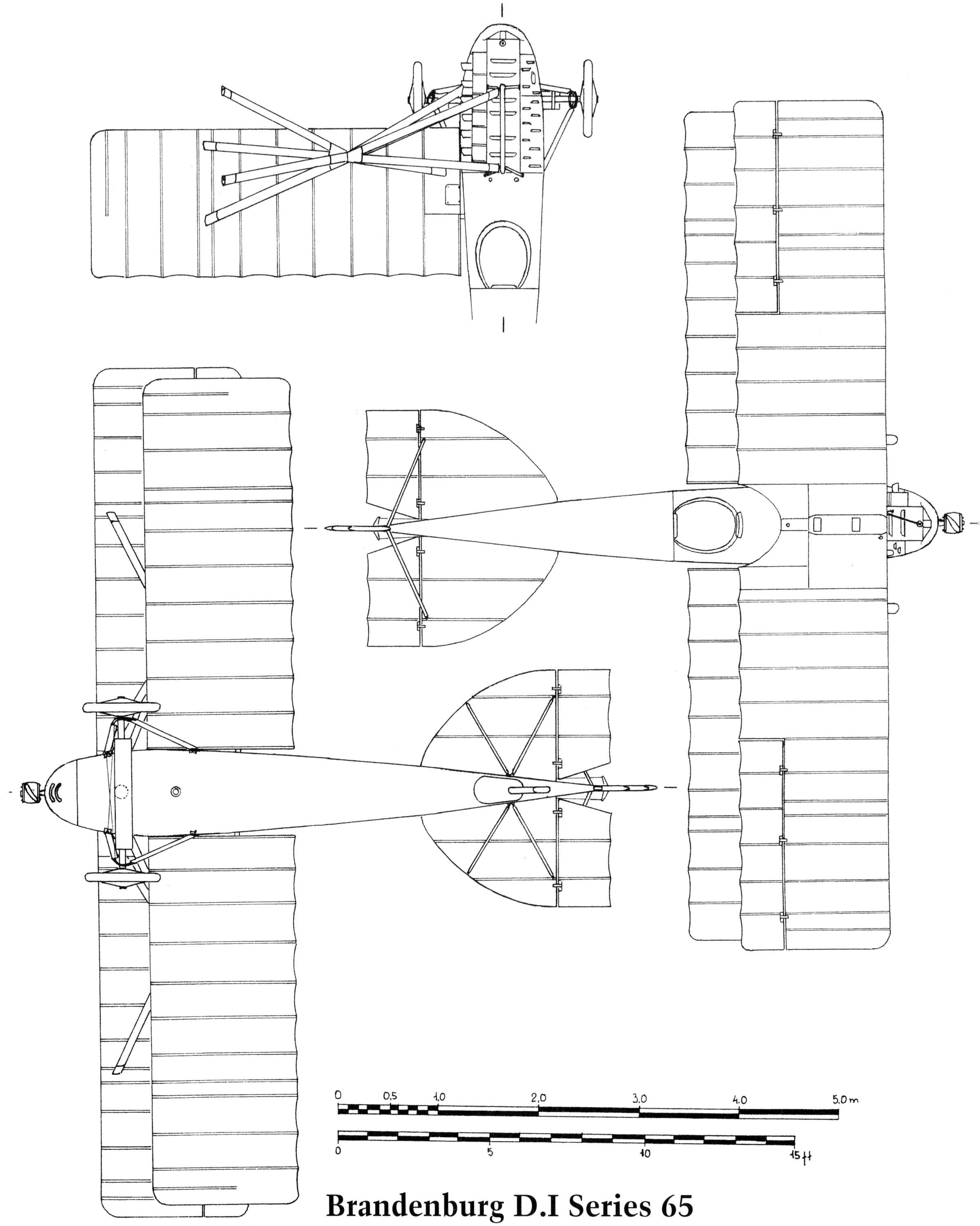

Brandenburg D.I Series 65

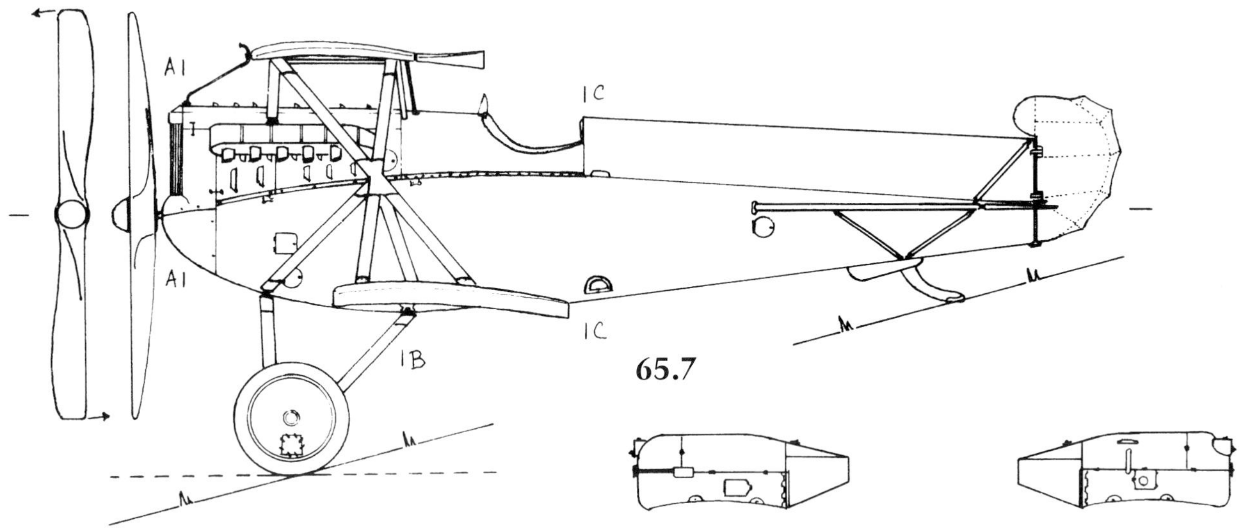

65.7

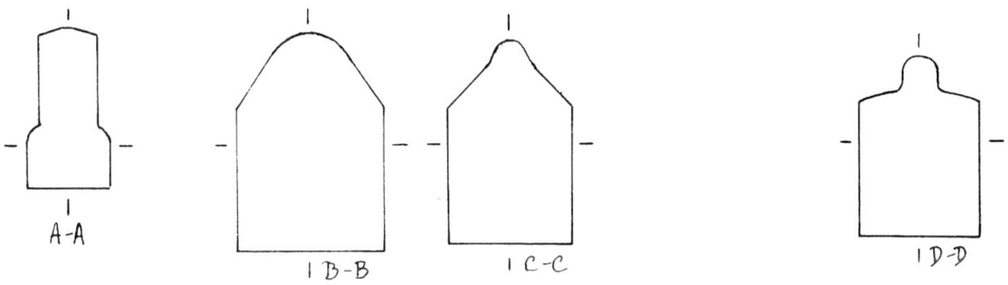

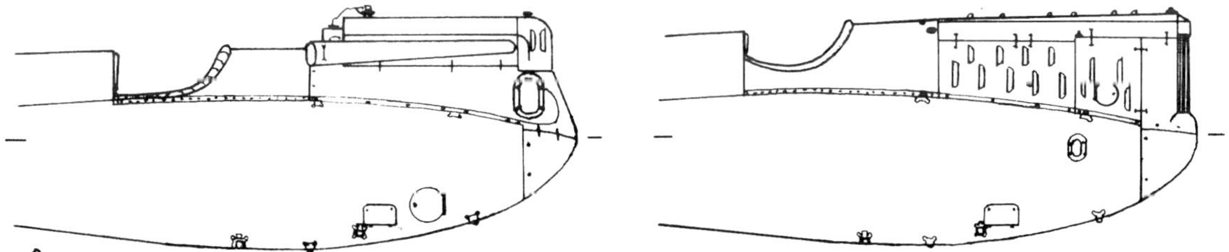

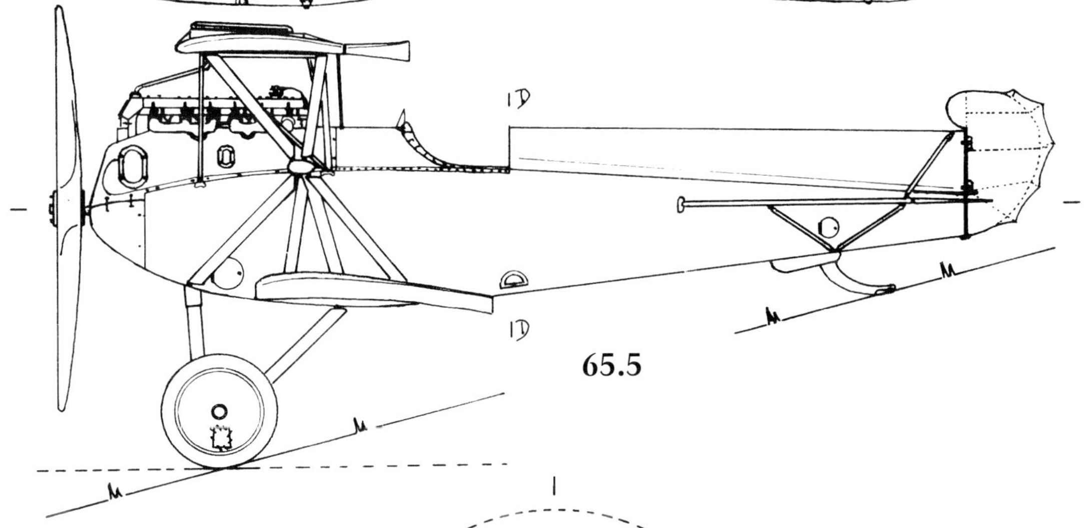

65.5

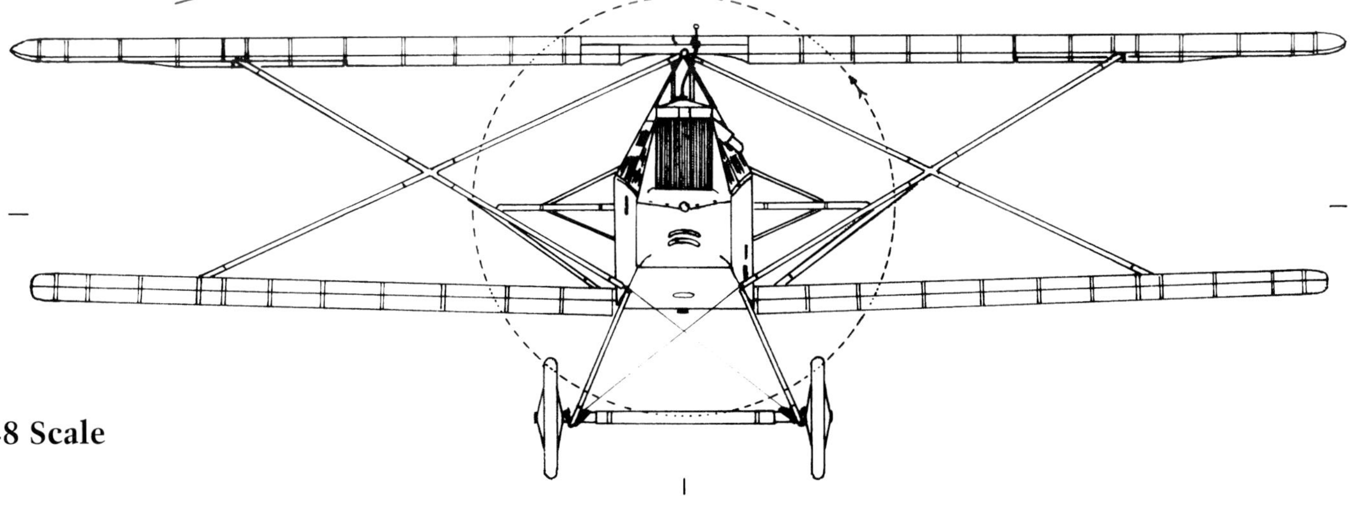

1/48 Scale

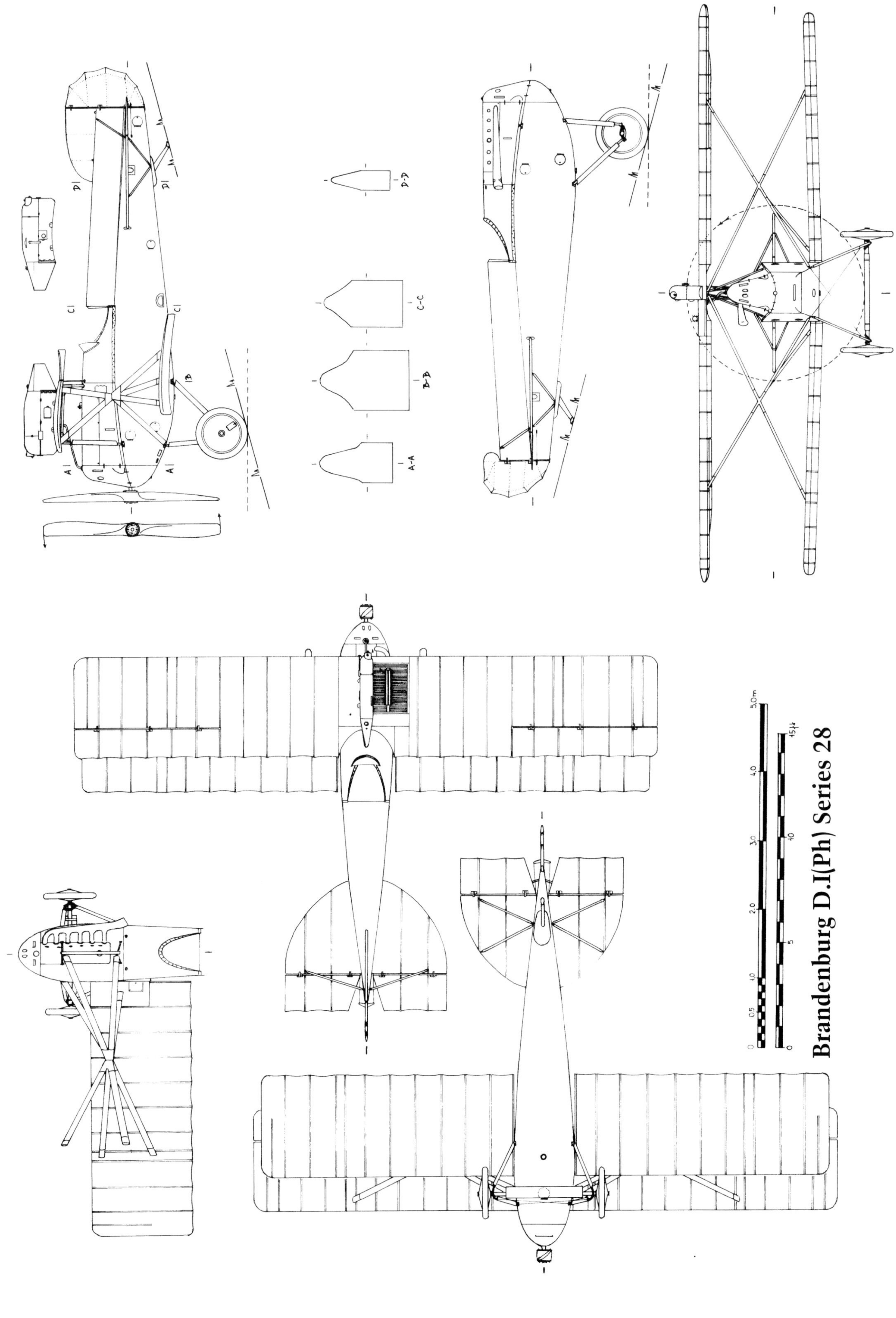

Brandenburg D.I(Ph) Series 28

1/72 Scale

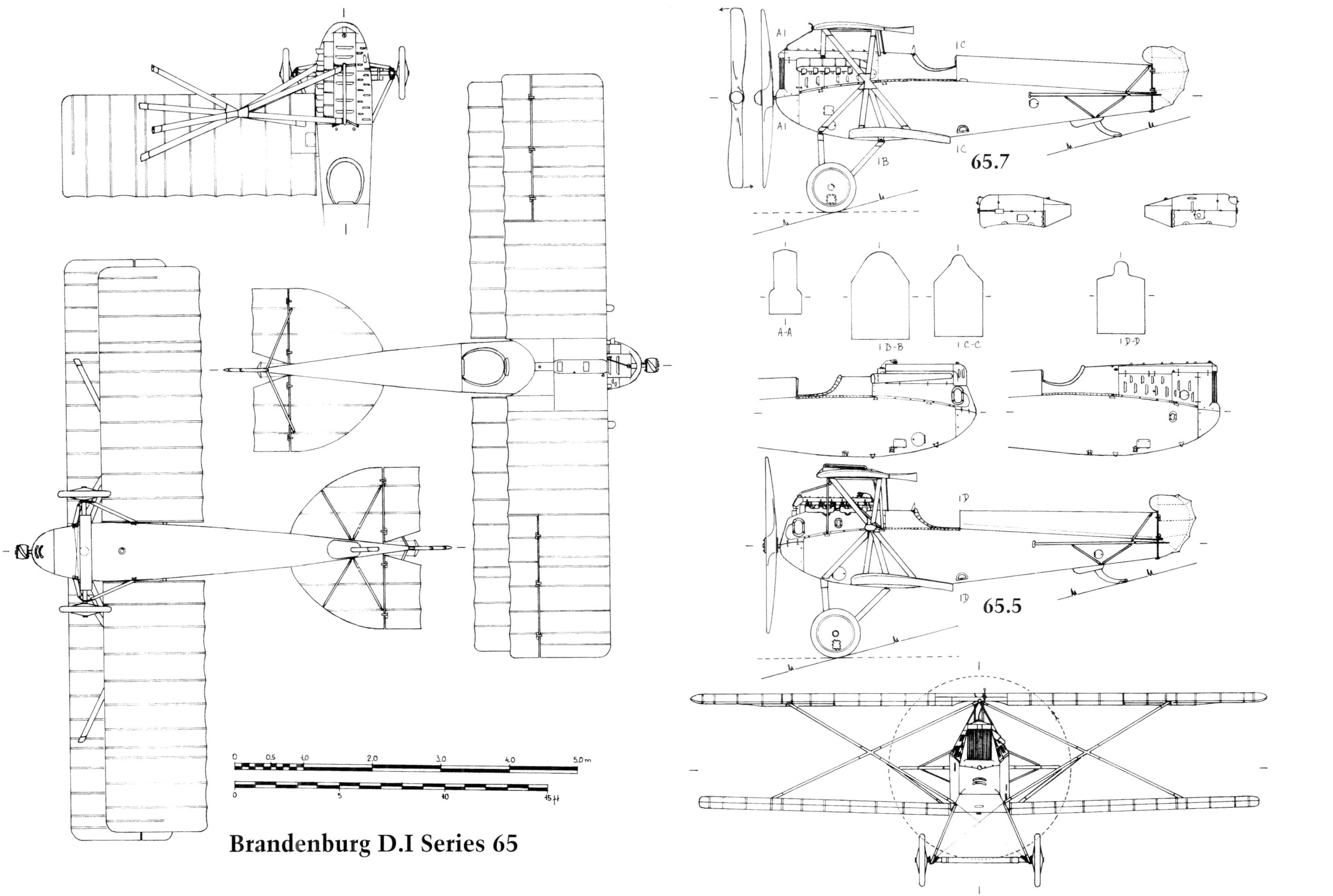

Brandenburg D.I Series 65

1/72 Scale

The cockpit of Brandenburg D.I 65.87.

Cockpit of Brandenburg D.I 65.87 with instruments identified.

Left and above: Two views of the crash of Brandenburg D.I 65.89.

Left, above left, below, and above: Four views of Brandenburg D.I 65.94.

Brandenburg D.I Details

Except for the photo at extreme upper left, photos on this page are of Brandenburg D.I 28.68. in the Prague Technical Museum. At the top right is a closeup of the propeller decal on propeller serial number 13.330.002.

Above: Left side of cockpit showing throttle linkage.

Above right: Instrument panel.

Right: Instrument panel.

Brandenburg D.I Details

All photos (except the two direcly below) Brandenburg D.I 28.68, work number 546, in the Prague Technical Museum.

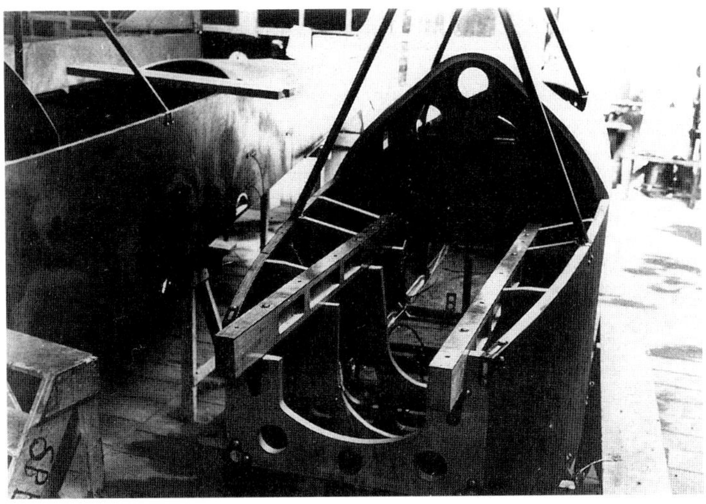

Brandenburg D.I in Combat

Foreword

By
Dr. Otto von Habsburg,
Son of the last Kaiser of Austro-Hungary, Karl I.

Shortly after the First World War, there appeared a book entitled *Armee im Schatten*[1], which enjoyed great success at the time. These words are indicative of the situation of the Austro-Hungarian Army during the four and one-half years of the First World War. To wit, much has been said as to the Germans and the French in the history of the war and, occasionally, about the English and the Americans, as well. But as to what Austro-Hungary achieved during this time, in general there has been silence, for following the destruction of the old Austria, there was no one to represent its interests in journalism.

In reality, a great deal had been accomplished, particularly from the perspective of the end of the Twentieth Century. Austro-Hungary had the only true multi-national army in which the various people of the monarchy jointly defended their country. And not least, to this army there belonged aviation, which had produced important technological advances as well.

Only today can we assess what happened many decades ago, concerned as we are with the question of security within the context of European unification and, along with that, we must inevitably come to the concept of a multi-national defense force. That it is possible for such a force to be successful has been demonstrated by the Austria of old. For this country, which had to defend itself against a world of enemies, to have stood up to this world of enemies for four and a half years, was an incredible accomplishment. The last great airman of this old army, Banfield, came from Triest.

It is also indicative that, upon the outbreak of the Second World War, the successor states to the Austro-Hungarian monarchy scarcely even put up a fight. The heavily-equipped Czech Army capitulated without firing a single shot. The Yugoslav Army, likewise one of the mightiest, disintegrated in less than four days because its people were not prepared to spill their blood for this country. But they had fought to the very end for Austro-Hungary.

I will never forget my meeting with the son of the Czech Republic's first president, Jan Masaryk. When we met in America and he was the foreign minister of the Benes government, he came up to me and said so that everyone could hear, "I would like you to know that I was no traitor. I was still fighting for Austria on the Fourth of November 1918." Moreover, the fact that he was one of the most highly decorated officers of the Imperial Austro-Hungarian Army was later concealed.

In this sense, it is good that journalism is now dealing with Austro-Hungarian aviation as well. It will show that mighty things were accomplished under the most difficult of circumstances. At a time when we wish to rebuild Europe, this demonstrates that a multi-national sense of community can develop which will prove its worth in difficult days as well.

Otto von Habsburg

[1] *Army in Shadow.*

Introduction: Why the KD Fighter?

When the terror of the Fokker single-seaters had vanished on the Western Front due to better Allied aircraft (especially the Nieuport 11), Germany's *Idflieg* (Inspectorate of the German Air Service) announced a competition for fighter aircraft which the Albatros Flugzeugwerke and its fighter airplane won hands-down. The Hansa-Brandenburg Flugzeugwerke, which had also taken part in the competition, found little favor in the strict eyes of the men of the test commission.

The rejected fighter prototype, which had been designed by Ernst Heinkel, was then offered to the Austro-Hungarian Ministry of War and, because there was no other choice available, it was accepted.

All of the KD models were extremely difficult to fly and this model was soon nicknamed "the killer" or "the flying coffin". Its armament was a problem in and of itself; in addition to which there then came the ban on the use of incendiary ammunition, which the Italians were using extensively.

Air Combat Victories

Italian Front – Isonzo 1916

Dec. 3: A Caproni bomber of Type Ca.2 (Number 1223) of *la Squadriglia da bombardemento, IV. Gruppo,* was shot down between 1300–1400 hours near Mavhinje (northeast of Triest) following bitter air combat. The machine burned upon impact with the ground.

Of the Caproni's four occupants, one man had been severely injured in the crash and subsequently died, one man was badly injured, and one man was slightly injured. Only the observer came out of the crash unscathed. Names and ranks of the occupants: *Tenente pilota* C. Mazzoni; *Tenente pilota* M. Borra (second pilot); *Tenente osservatore* L. Guzzanti (uninjured); *Caporale maggiore* O. Castoldi, machine-gunner.

The above air victory was achieved jointly by a combination of the following Austro-Hungarian naval and army pilots: The first attack was by *FP Zugsf.* Karl Cislaghi of *Flik* 28 flying KD 65.67 (armament consisted of a Schwarzlose MG 7/12). This was *Zugsf.* Cislaghi's second air victory, for which he was decorated with the Golden Bravery Medal.

The second attacker was *Linienschiffsleutnant* Gottfried Banfield, commander of the Triest Naval Air Station, with naval KD "A.12" (formerly designated "CC"). According to the statement made by the captured Italian observation officer, Banfield deserved most of the credit for this victory.

After Banfield broke off the attack because of a machine-gun jam he couldn't clear, Hansa-Brandenburg C.I 20.08 of *Flik* 19 with *FP Feldw.* Heinrich Mahner (first air victory) and *BO Lt.* Alexander Tahy (first air victory) attacked the Caproni in an extremely daring fashion. In so doing, Lt. Tahy was slightly wounded by a grazing shot to the hip and his own airplane was badly shot up.

Feldw. Mahner was nominated for the Golden Bravery Medal, however, he received the Silver Bravery Medal, 1st Class, for the second time, instead. *Lt.* Tahy received the Bronze Military Service Medal with Swords.

The Caproni was given the *coup de grâce* by its final

attacker, *FP Oblt.* Godwin Brumowski of *Flik* 12, who had taken off for this defensive patrol in KD 65.53 (with a Schwarzlose MG 7/12).

Oblt. Brumowski and *Linienschiffsleutnant* Banfield were mentioned by name in the Austro-Hungarian Army Report of 5 Dec. 1916. *Linienschiffsleutnant* Banfield was also mentioned by name in the Austro-Hungarian Fleet Report of 6 Dec. 1916.

The downing of the heavily-armed Caproni tri-motor (the first "flying fortress") has always been acknowledged as an extraordinary achievement.

Postscript to Brumowski: his fourth air victory.

Italian Front – Isonzo 1917

Almost all of the downed aircraft went down behind the Italian lines or in the hinterland. Aircraft crashing in Austrian territory are given special note. The Isonzo front is understood to mean the combat region between the North Adriatic Sea and Flitsch/Tolmein.

Feb. 13: Farman reconnaissance plane, crashed between the lines south of Kanstanjevica (east of Gorizia) at approximately 1700 hours. Confirmed by *Stoluft der k.u.k 5. Armee, Oberst der Artilleriestabes* Leo Nadherny. Fifth air victory of *FP Hptm.* Raoul Stojsavljevic, assigned to *Flik* 34 as a fighter pilot, flying KD 65.68.

Feb. 25: Nieuport single-seater, forced to land in the Cervignano—Aquileja area, during an escort flight for Hansa-Brandenburg C.I 29.53 (*Korp.* Slanina and *BO Oblt.i.d.Res.* von Heintschel). Second air victory of *FP Feldw.* Ernst Franz of *Flik* 23, flying KD 65.63. For this and other successful cover flights, the pilot received the Golden Bravery Medal.

In the monthly report of *Flik* 23 for February 1917, the air combats are indeed cited, however, an air victory is not mentioned. Hence, that the enemy had gone down must have been confirmed later by reports from the front. Because of his courageous and skillful conduct, the KD pilot prevented a new reconnaissance plane being shot down, saved a brave crew from injury or death, and *Flik* 23 retained the only usable camera the company possessed, as well.

April 13: Nieuport single-seater, crashed, cartwheeling several times. Because the victory occurred under very poor visual conditions far to the rear in the Italian hinterland, the pilot, *FP Oblt.* Frank Linke-Crawford of *Flik* 12, forwent both a victory report and the confirmation process.

In a letter to his sister, Eileen, Linke-Crawford mentions that the Italian pilot was clearly a rank beginner who ran away out of fear the moment he noticed Crawford's KD 28.02 with the "red nose".

As a interesting aside, it should be noted that Linke-Crawford was already using a personal emblem at *Flik* 12: the forward part of 28.02's engine was painted red.

April 24: Voisin reconnaissance plane, forced to land near Farra (south of Gorizia), confirmed by *Flugwache* II (Observation Post II). The eighth air victory of *FP Offstv.* Julius Arigi of *FlG* I, flying KD 28.06.

May 3: Farman reconnaissance plane, crashed at approximately 1045 hours near Nogaredo (between Cormons and Palmanova). Confirmed by *Lt.* Noha, commander of *Flugwachen* I and II, as well as by photo reports of the crew, *Korp.* Hugo König and *BO Lt.i.d.Res.* Ernst Pirnos, of *FlG* I. Ninth air victory of *Offstv.* Julius Arigi of *FlG* I, flying KD 28.06.

After Arigi returned, the aforementioned crew took off with Hansa-Brandenburg C.I 129.39 (129.38, according to other sources) to take a photograph of the downed airplane. Attacked over the front by a Nieuport single-seater, the crew succeeded in confirming that the enemy machine had been brought down (an emergency landing north of Gradisca). The photograph was taken later.

Feldpilot Offizierstellvertreter Josef Kiss, 19 air victories, KIA 24 May 1918. Photo: Lahner Archive via Karl Meindl.

May 4: Farman reconnaissance airplane, crashed at approximately 1700 hours, over the Podgora plateau and was immediately thereafter set on fire by artillery of the *k.u.k. 58. Feldartillerie-Brigade.* Confirmed by artillery reports and by *Stoluft der k.u.k. 5. Armee.* Tenth air victory of *FP Offstv.* Julius Arigi of *FlG* I flying KD 28.06. (Podgora is a plateau north of Gorizia which was the scene of heavy fighting up until the Sixth Battle of the Isonzo.)

May 10: Voisin reconnaissance airplane, forced to land near Monfalcone, confirmed by *Stoluft der k.u.k. 5. Armee.* Sixth air victory of *FP Hptm.* Godwin Brumowski, commander of *Flik* 41/J, flying KD 28.10.

May 12: Farman reconnaissance airplane, crashed near Podsenica (north of Gorizia), confirmed by reports from the front. Fourth air victory of *FP Feldw.* Karl Kaszala of *Flik* 41/J, KD number unknown.

May 12: Farman reconnaissance plane, went down in a dive west of Gorizia, confirmed by reports from the front. Seventh air victory of *FP Hptm.* Godwin Brumowski, commander of *Flik* 41/J, flying KD 28.10.

According to Italian sources, a Farman of *27a Squadriglia da ricognizione* was shot down in air combat on the Isonzo front on May 12, 1917. The crew members, *Tenente pilota* Luigi Giovanardi and *Capitano osservatore* Armando Giua, were uninjured.

The question as to who was the victor, Kaszala or Brumowski, can no longer be determined.

May 12: Nieuport reconnaissance airplane, forced into a steep descent, west of Gorizia. *Hptm.* Godwin Brumowski, flying

Feldpilot Offizierstellvertreter Karl Kaszala, 8 air victories, killed 1936.

Feldpilot Feldwebel Hermann Richter, a KD pilot, *Flik* 41/J. Photo: Dr. Reinhard Masajdek.

KD 28.10, and *Feldw.* Karl Kaszala, flying an unknown KD, both took part in this unconfirmed victory.

The sources are quite muddled. In the activity report for May 1917, a Farman is reported as confirmed; in Brumowski's combat flight log, however, a Nieuport is listed. Kaszala's participation is evident from the activity report.

May 13: A two-engine Voisin reconnaissance airplane (Caudron G.4?), forced to land in the area of Plava, unconfirmed. Claimed as an air victory by *FP Hptm.* Godwin Brumowski, flying KD 28.11, and *FP Feldw.* Jaroslaus Samek, flying KD 65.65, both of *Flik* 41/J. An obvious error in type designation, as the Voisin was a single-engine airplane.

May 14: Farman reconnaissance airplane, crashed smoking near Gradiscutta (northwest of Gorizia), confirmed by *Stoluft der k.u.k. 5. Armee.* Eleventh air victory of *FP Offstv.* Julius Arigi of *FlG* I, flying KD 28.08. Arigi was mentioned by name in the Austro-Hungarian Army Report of 16 May 1917.

May 15: Voisin reconnaissance airplane, crashed between San Canziano and Bestrigna (south of Gorizia). Confirmed by *Stoluft der k.u.k. 5. Armee.* Twelfth air victory of *FP Offstv.* Julius Arigi, flying KD 28.08.

For his eleventh and twelfth air victories, Arigi was decorated with the Silver Bravery Medal, First Class, for the second time.

May 20: SPAD two-seater (with French cockades), shot down over Monte Santo (north of Gorizia) at approximately 1035 hours. Confirmed by *Stoluft der k.u.k. 5. Armee.* Eighth air victory of *FP Hptm.* Godwin Brumowski, flying KD 28.10. Fifth air victory of *FP Feldw.* Karl Kaszala, flying KD 28.11, both of *Flik* 41/J. (The cover painting is of this action.)

June 25: Nieuport single-seater, unconfirmed air victory of *FP Oblt.* Frank Linke-Crawford of *Flik* 12, flying KD 28.38.

June 30: Nieuport single-seater, crashed over the Podgora plateau at approximately 1000 hours, confirmed by two *Flugwachen* and by *Oblt.i.d.Res. FP* Ludwig Hautzmayer of *Flik* 19. First air victory of *Feldw.* Johann Risztics of *Flik* 42/J, flying KD 28.23. The victorious pilot was decorated for the second time with the Silver Bravery Medal, First Class.

July 5: Caudron reconnaissance airplane, crashed near Ronchi. Confirmed by infantry reports, two *Flugwachen* and two separate airplane crews. Second air victory of *FP Feldw.* Vinzenz Magerl of *Flik* 42/J, flying KD 28.18.

July 10: SPAD and Nieuport single-seaters, crashed over Monte Sabotino at approximately 1800 hours, the Nieuport was destroyed by Austro-Hungarian artillery fire. Confirmed by *Flugwache* reports, artillery, and infantry. First and second air victories of *FP Zugsf.* Ferdinand Udvardy of *Flik* 42/J, flying KD 28.42.

The pilot rescued two reconnaissance aircraft of *Flik* 2 with his daredevil attack on the enemy fighters; for this he was decorated with the Gold Bravery Medal.

July 17: A twin-engine Voisin reconnaissance airplane, crashed in flames in the Isonzo river between Monte Sabotino and Monte Santo and was smashed to pieces. Confirmed by *Flugwache* II, the *58. Feldartillerie-Brigade,* and the *24. Infanterie-Division.* Ninth air victory of *FP Hptm.* Godwin Brumowski, flying KD 28.57; second air victory of *FP Zugsf.* Ferdinand Jaschek, flying KD 28.44; second air victory of *FP Zugsf.* Josef Novák, flying KD 28.33, all of *Flik* 41/J.

An obvious error as to type, the aircraft was a Farman

Feldpilot Oberleutnant Frank Linke-Crawford, "The Falcon of Feltre", 27 air victories, KIA 31 July 1918. Photo: Karl Meindl.

Feldpilot Feldwebel Karl Sailer, KIA 14 July 1917. Photo: Karl Meindl.

reconnaissance airplane of 45a *Squadriglia da ricognizione*, according to Italian sources. The crew, *Tenente pilota* Leo Vidotti and *Tenente osservatore* Fernando Silvestri, were wounded in air combat. Time of the victory was approximately 1100 hours. *Zugsf.* Jaschek, to whom most of the credit for the victory was due, was decorated with the Gold Bravery Medal.

Aug. 9: Nieuport single-seater, crashed at approximately 0830 hours in the Auzza-Plava-Canale area (Kolowrat). Confirmed by an escort airplane (reconnaissance), *Flik* 42/J, and *Stoluft der Isonzo Armee.* Third air victory of *FP Oblt.i.d.Res.* Benno Fiala *Ritter* von Fernbrugg of *Flik* 12, flying KD 28.38, and third air victory of *FP Feldw.* Johann Risztics of *Flik* 42/J, KD number unknown.

Aug. 10: Nieuport single-seater, forced to descend with a dead engine in the area of Chiapovano (south of Gorizia), later confirmed by the *XV. Korpskommando* (Corps Headquarters). Tenth air victory of *FP Hptm.* Godwin Brumowski, commander of *Flik* 41/J, flying KD 28.69.

Aug. 10: Caproni bomber, shot down at approximately 1930 hours in the Plava–Auzza area, confirmed by *Abschnitts-kommando* IIa (Sector Headquarters). Fourth air victory of *FP Oblt.i.d.Res.* Benno Fiala *Ritter* von Fernbrugg of *Flik* 12, flying 28.38.

Aug. 11: Caudron reconnaissance airplane, crashed in flames during the morning hours near Plava. Confirmed by infantry reports, *Abschnittskommando* IIa, as well as by the *XV. Korpskommando.* Eleventh air victory of *FP Hptm.* Godwin Brumowski, flying KD 28.57; third air victory of *FP Zugsf.* Ferdinand Jaschek, flying an unknown KD, both of *Flik* 41/J.

Aug. 11: SPAD single-seater, shot down over the Hermada plateau, at approximately 1800 hours. Fourth air victory of *FP Feldw.* Johann Risztics of *Flik* 42/J, flying an unknown KD. For his victories on 9 and 11 August 1917, Risztics received the Gold Bravery Medal.

Aug. 11: SAML reconnaissance airplane, crashed in flames near Plava at approximately 1930 hours. Confirmed by *Abschnittskommando* IIa and by a *k.u.k. Schützen-Division* (Rifle Division). Fifth air victory of *FP Oblt.i.d.Res.* Benno Fiala *Ritter* von Fernbrugg of *Flik* 12, flying KD 28.38. For his confirmed victories on 9, 10, and 11 August, 1917, Fiala-Fernbrugg received the Order of the Iron Crown, Third Class, with War Decoration and Swords.

Aug. 11: Caudron reconnaissance airplane, crashed in flames west of Plava at approximately 2000 hours; enthusiastically confirmed by the Austro-Hungarian infantry. Twelfth air victory of *FP Hptm.* Godwin Brumowski, commander of *Flik* 41/J, flying KD 28.69. For his three air victories, *Hptm.* Brumowski was mentioned by name in the Austro-Hungarian Army Report of 13 August 1917. KD pilots Brumowski, Fiala-Fernbrugg, and Risztics received commendatory recognition from *General der Infanterie* Johann *Ritter* von Henriquez for their air victories on 11 August 1917.

As for the reconnaissance aircraft downed on 11 August 1917, the type designations here cannot be correct, either. Italian sources report total losses for 11 August 1917, as being one Farman reconnaissance plane of *45a Squadriglia da*

Feldpilot Korporal Heinrich Mayrbäuerl, KIA 20 July 1918, 3 air victories. Photo: Austrian War Archive.

A group of Austro-Hungarian aces, from left to right: *Feldpilot Offizierstellvertreter* Julius Kowalczik, *Feldpilot Offizierstellvertreter* Josef Kiss, and *Feldpilot Offizierstellvertreter* Karl Kaszala, Spring 1918, Budapest. Photo: Franz Selinger.

ricognizione. The crew members, *Sergente pilota* Melchiore Cigna and *Tenente osservatore* Francesco Allegra, were killed in air combat. Since no details as to the time and place of the crash are given in the Italian accounts, it is impossible to classify this loss as to the Austro-Hungarian pilot involved.

Aug. 12: Caudron reconnaissance airplane, crashed in flames near Selo, unconfirmed. Claimed by *FP Oblt.i.d.Res.* Benno Fiala *Ritter* von Fernbrugg of *Flik* 12, flying KD 28.38.

Aug. 14: Nieuport single-seater, crashed south of Gorizia at approximately 0630 hours. Confirmed by *Abschnittskommando* IIa and *Stoluft der Isonzo Armee.* Sixth air victory of *FP Oblt.i.d.Res.* Benno Fiala *Ritter* von Fernbrugg of *Flik* 12, flying KD 28.66.

Aug. 14: Two Nieuport single-seaters, shot down over Fajti Hrib and near Villa Vicentina (south of Gorizia). Confirmed by *Flugwache* reports, *Flik* 23, as well as by *Stoluft der Isonzo Armee.* First and second air victories of *FP Feldw.* Alphons Korsitzky of *Flik* 28, flying KD 28.28. For this remarkable deed, Korsitzky was decorated with the Gold Bravery Medal. The plane he used, KD 28.28, was on the rolls of *Flik* 23 and had been "loaned out" for this combat flight!

Aug. 14: Two Macchi M5s—seaplane fighters. One seaplane shot down in flames near Grado, confirmed by infantry reports; one seaplane forced to land, unconfirmed. Thirteenth air victory of *FP Hptm.* Godwin Brumowski, commander of *Flik* 41/J, flying KD 28.69.

Described as a "Water Nieuport" in Brumowski's combat flight log. The many erroneous type descriptions of downed aircraft on both sides, Austrian as well as Italian, is a problem for any conscientious researcher. In view of such vague sources, assigning Italian losses so as to fit the Austrian air victories is possible only with the greatest skepticism.

Aug. 18: Two Caudron reconnaissance aircraft, forced to land over Monte Santo; one air victory confirmed through infantry reports. Fourteenth air victory of *FP Hptm.* Godwin Brumowski, commander of *Flik* 41/J, flying KD 28.69.

Also an erroneous type description. According to Italian accounts, it was either a Farman or an S.P. reconnaissance airplane of 27a *Squadriglia da ricognizione.* (This squadron's equipment was a mix of Farman and Savoia-Pomilio two-seaters.) The crew members, *Tenente pilota* Achille Pierro and *Capitano osservatore* Ottorino Mutti, were uninjured during the forced landing.

Aug. 19: Caproni bomber, shot down in flames north of Fajti Hrib–Plava during the morning hours. Confirmed by the *VII. Korpskommando* and *Stoluft der Isonzo Armee.* Seventh air victory of *Oblt.i.d.Res.* Benno Fiala *Ritter* von Fernbrugg of *Flik* 12, flying KD 28.38.

According to Italian sources, possibly a Caproni of 5a *Squadriglia da bombardemento.* Of the crew, the two pilots, *Tenente pilota* Guiseppe Orselli and *Tenente pilota* Pierino Gallia, are said to have escaped with their lives. As for the rest of the crew, only their ranks and names are known (*Tenente osservatore* F. dal Passano, *Sergente mitragliere* G. Moriano); what became of them is unknown.

Aug. 19: SPAD single-seater, shot down near Rubbia, confirmed by *Flugwache* reports and by *Stoluft der Isonzo Armee.* Third air victory of *FP Feldw.* Alphons Korsitzky of *Flik* 28, flying KD 28.59. In further air combat with enemy fighter aircraft, the pilot had to make an emergency landing behind his own lines on account of some 40 hits (his fuel tank being shot through, among other things). The airplane was badly damaged and was written off in September 1917. Korsitszky was uninjured.

Aug. 19: Two Macchi M5 seaplane fighters forced to land, neither victory confirmed. Claimed by *FP Hptm.* Godwin Brumowski, commander of *Flik* 41/J, flying 28.69.

Aug. 20: Caudron reconnaissance airplane, shot down in the area of San Giovanni–Manfalcone, confirmed by *Stoluft der Isonzo Armee.* Sixteenth air victory of *FP Hptm.* Godwin Brumowski, flying Albatros D. III 153.06; first air victory of *FP Feldw.* Hermann Richter, flying KD 28.40; first air victory of *Korp.* Heinrich Mayrbäurl, flying KD 28.65, all of *Flik* 41/J.

Aug. 20: Caudron reconnaissance airplane, shot down near Vertojba (south of Gorizia), confirmed by infantry reports. Seventeenth air victory of *FP Hptm.* Godwin Brumowski, flying KD 28.69, and the fourth air victory of *FP Zugsf.* Ferdinand Jaschek, flying an unknown KD, both of *Flik* 41/J.

Again, an erroneous type designation for both of the air victories on 20 August 1917. According to Italian sources, this

Feldpilot Hauptmann Franz Rabitsch, CO of *Flik* 9 and *Fluggeschwader* I, killed 1 March 1917. Photo: Josef Bangerl via Karl Meindl.

Above: *Feldpilot Feldwebel* Erwin Funk, *Flik* 16 and 41/J. Photo: Karl Meindl.

Right: *Feldpilot Oberleutnant* Ludwig Hautzmayer, 7 air victories, *Flik* 19, 51/J, and CO of *Flik* 61/J. Photo: Dr. Werner Mörath via Karl Meindl.

Left: *Feldpilot Leutnant* Richard Melzer, *Flik* 42/J. Photo: *Heeresgeschichtliches Museum* (Army History Museum), Vienna.

was actually an S.P. reconnaissance plane of 35a *Squadriglia da ricognizione;* the aircraft was badly damaged. The crew, *Sergente pilota* P. Fusar and *Tenente osservatore* Umberto Nistri, managed to escape with their lives.

Aug. 21: Nieuport single-seater, crashed over Monte Santo, confirmed by *Flugwache* reports. Eighteenth air victory of *FP Hptm.* Godwin Brumowski, flying KD 28.69; first air victory of *FP Oblt.* Frank Linke-Crawford, flying KD 28.40, both of *Flik* 41/J.

Aug. 22: S.P. reconnaissance airplane made an emergency landing on the Gorizia drill ground, was shot to pieces by

Austro-Hungarian artillery and confirmed by artillery reports. Nineteenth air victory of *FP Hptm.* Godwin Brumowski, flying KD 28.69; second air victory of *FP Oblt.* Frank Linke-Crawford, flying KD 28.40, both of *Flik* 41/J. (According to the combat flight logs of Brumowski and Linke-Crawford, it was a Caudron; according to all other sources, it was a Savoia-Pomilio two-seater.)

Aug. 23: S.P. reconnaissance airplane, forced to land in its own territory near Bate (Bainsizza Plateau) at 1040 hours. Airplane was supposedly captured with little damage, then destroyed by the Austro-Hungarian infantry during the retreat.

The crew members, *Sergente pilota* Ermanno Malaspina and *Tenente osservatore* Cesare Mazzarino of the *45a Squadriglia da ricognizione,* were taken prisoner unharmed. After being taken prisoner, the pilot and observer were taken to *Flik* 41/J at Sesana and were well fed, as many photographs prove. Both of the Italians were delighted to have escaped with their lives and that the war was over for them.

For his eighteenth air victory, *Hptm.* Brumowski was mentioned by name in the Austro-Hungarian Army Report of 26 August 1917. (As can be seen, the rate at which Brumowski was shooting down airplanes was much faster than the laborious and very bureaucratic confirmation process.)

For his many successful combat flights with the KD type, *Hptm.* Brumowski was recommended for the Knight's Cross of the Order of Leopold with War Decoration and Swords by the *Stabsoffizier der k.u.k. Isonzo Armee, Oberst des Artilleriestabes* Leo Nadherny, on 12 August 1917. In his first statement, *Generaloberst* Svetozar Boroevic von Bojna, commander of the *Isonzo Armee,* pleaded for waiting, even though the application had also been warmly recommended to the *Kommandant der k.u.k. Luftfahrtruppen* (Commander of the Austro-Hungarian Air Service) by *Oberst* Emil Uzelac. Only after the indubitable air victory of 23 August 1917 (see above), did Boroevic give his unreserved recommendation for the award in a telegram, with the following words of recognition: "...particularly splendid, brave achievements..."

Twentieth air victory (officially, the eighteenth air victory) of *FP Hptm.* Godwin Brumowski, flying KD 28.69; third air victory of *FP Oblt.* Frank Linke-Crawford, flying KD 28.40; second air victory of *Korp.* Heinrich Mayrbäurl, flying KD 28.65, all of *Flik* 41/J.

Aug. 24: Nieuport single-seater, shot down in the area of Gorizia, confirmed by *Flugwache* reports. First air victory of *FP Feldw.* Gottlieb Czernin of *Flik* 23, flying an unknown KD. Voluntary escort flight for *Flik* 46/F and *FlG* I, which were bombarding the Italian positions and troop concentrations in the area of Plava on this particular day.

Aug. 25: Sopwith two-seater, forced into a steep descent after several attacks. During the last attack, the observer was lying motionless, either dead or severely wounded, across his observer's machine-gun; victory unconfirmed. Claimed as an air victory by *FP Hptm.* Godwin Brumowski, flying KD 28.69, and *FP Oblt.* Frank Linke-Crawford, flying KD 28.40, both of *Flik* 41/J.

This too is an erroneous type designation (in Linke-Crawford's diary, it is described as a "lattice-tail", in the combat flight logs as a "Sopwith two-seater", hence referred to as a "fuselaged biplane").

Aug. 25: S.P. reconnaissance airplane, forced into a steep descent in the area of Cepovan, unconfirmed. Claimed as an air victory by *FP Hptm.* Godwin Brumowski, flying KD 28.69, and *FP Oblt.* Frank Linke-Crawford, flying KD 28.40, both of *Flik* 41/J.

Aug. 26: SPAD single-seater, severely damaged following a successful attack and forced into a steep descent during an

Above: *Feldpilot Hauptmann* Godwin Brumowski, *Fliks* 1, 12, and CO of *Flik* 41/J, 35 air victories. Photo: Oswald Kostrba.

Right: *Feldpilot Linienschiffsleutnant* Gottfried *Freiherr* von Banfield, the leading Austro-Hungarian naval ace, 9 air victories. Photo: Karl Meindl.

Left: *Feldpilot Hauptmann* Godwin Brumowski, CO of *Flik* 41/J, 35 air victories, August 1917. Photo: *Die Leipziger Illustrierte.*

Right: *Feldpilot Oberleutnant* Benno Fiala *Ritter* von Fernbrugg, *Fliks* 1, 41/J, 12 and CO of *Flik* 51/J, 28 air victories. Photo: Karl Meindl.

Left: *Feldpilot Offizierstellvertreter* Josef Siegel, *Fliks* 15, 21, 6, 16, and *Flik* 56/J. Photo: Dr. Reinhard Masajdek, Vienna.

escort flight over Fajti Hrib (a bombing raid by *Flik* 46/F on Italian troop concentrations); confirmed by infantry reports. Second air victory of *FP Hptm.i.d.Res.* Karl Banfield, commander of *Flik* 46/F, flying Albatros D.II 53.13, and *FP Feldw.* Gottlieb Czernin, flying an unknown KD (also his second victory).

According to other sources, it was probably a Nieuport. For his voluntary participation in the missions of *Flik* 46/F (see also the air victory of 24 August 1917), *Feldpilot* Czernin was recommended for the Bronze Bravery Medal by *Hptm.i.d.Res.* *FP* Banfield, an unusual occurrence, as Czernin was a member of *Flik* 23.

It is certain that Czernin was flying a KD during these escort flights. Unfortunately, there are no combat flight logs for this pilot. It is possible that he was either using a KD from *Flik* 23 (28.05 or 28.28) or that he had been loaned the single KD still on hand at *Flik* 46/F, namely KD 28.62, for these escorts flights.

Aug. 28: S.P. reconnaissance airplane, forced to land, unconfirmed. Claimed as an air victory by *FP Hptm.* Godwin Brumowski, commander of *Flik* 41/J, flying KD 28.69.

The Austro-Hungarian confirmation process for recognizing air victories was fairly involved and bureaucratic, but very stringent. The above air victory by *FP Hptm.* Godwin Brumowski had not been confirmed by front-line reports. The above air victory was the only aircraft brought down on the Isonzo front on this particular day.

According to Italian accounts, on this day, an S.P.3 reconnaissance airplane of *22a Squadriglia da ricognizione* was lost (total loss) as a result of air combat. The crew, *Sergente pilota* Donato Ernesti and *Tenente osservatore* Antonio Picentino, either fell in air combat or were killed in the crash.

Oct. 25: SPAD single-seater, crashed in flames in Austrian territory south of Monte San Gabriele, confirmed by infantry reports. Eighth air victory of *Oblt.i.d.Res.* Benno Fiala *Ritter* von Fernbrugg of *Flik* 12/D, flying KD 28.38. For his sixth through eighth confirmed air victories, Fiala-Fernbrugg was awarded the Knight's Cross of the Order of Leopold with War Decoration and Swords.

Italian Front – South Tyrol 1917

June 10: Nieuport single-seater, crashed just behind the Italian positions near Asiago, destroyed as a result of shelling by the Austro-Hungarian artillery. Confirmed as the fourth air

Feldpilot Hauptmann Karl Nikitsch, *Flik* 16, CO of *Flik* 14, CO of *Fliks* 39 and 63/J, 6 air victories. Photo: War Archive, Vienna.

Feldpilot Oberleutnant Ernst Strohschneider, *Flik* 23, *Fluggeschwader* I, *Flik* 42/J, CO of *Flik* 61/J, 15 victories, killed 21 March 1918. Photo: Karl Meindl.

Feldpilot Offizierstellvertreter Kurt Gruber, *Flik* 1, *Fliks* 41/J and 60/J, 11 air victories, KIA 4 April 1918. Photo: War Archive, Vienna.

Feldpilot Offizierstellvertreter Julius Arigi, *Flik* 6, *Fluggeschwader* I, *Flik* 55/J, *Flik* 6/F. Photo: Not attributed.

victory of *FP Feldw.* Josef Kiss of *Flik* 24 by artillery reports as well as by the *Stabsoffizier der k.u.k. Luftfahrtruppen* with the *11. Armee, Rittm. FP* Wilhelm Weldin. Kiss was flying KD 28.37, which had a synchronized machine-gun with the Bernatzik synchronization mechanism.

June 14: SAML reconnaissance airplane of *113a Squadriglia da ricognizione,* forced to land in Austrian territory between Roana and Lavarone. The crew, *Aspirante ufficiale pilota* Filippo Rossi and *Tenente osservatore* Guiseppe Negro, were taken prisoner.

Fifth air victory of *FP Feldw.* Josef Kiss of *Flik* 24, flying KD 28.37. For his fourth and fifth victories, Kiss was decorated for the second time with the Gold Bravery Medal, in addition to being favored with a promotion to *Stabsfeldwebel* on 1 July 1917.

July 13: S.P.3. reconnaissance airplane (with the number "4530") of 26a *Squadriglia da ricognizione,* forced to land in Austrian territory near the area of Levico in the Sugana Valley. By making his daring landing next to the Savoia-Pomilio, *Stabsfeldwebel* Kiss prevented the enemy machine from being destroyed, and the airplane was captured almost intact (this was the first S.P. reconnaissance plane to fall into Austro-Hungarian hands).

The pilot, *Tenente pilota* Vitale Piga-Spinas, was taken prisoner, the observer, *Tenente osservatore* Renato Semplicini, was killed in air combat and was buried with full military honors.

This victory was achieved jointly by a combination of the following Austro-Hungarian airmen: Sixth air victory of *Stabsfeldwebel* Josef Kiss of *Flik* 24, flying KD 28.37, who can claim the greater part of this success and who was honored by being mentioned by name in the Austro-Hungarian Army Report of 15 July 1917. First air victory of *FP Feldw.* Viktor Zimmermann of *Flik* 48, with Albatros D. III 53.33. Fourth air victory of *FP Feldw.* Karl Maurer and first air victory of *BO Oblt.* Erich Kühne of *Flik* 17, flying Hansa-Brandenburg C.I 29.68.

Sep. 14: Italian fuselaged biplane, forced to land in the Arsiero area as a result of engine trouble (radiator shot to pieces),

NCO pilots of *Fluggeschwader* I being decorated by their commanding officer, *Hauptmann* Karl Sabeditsch, Divacca airfield, Summer 1917. Photo: Josef Bangerl via Karl Meindl.

according to later artillery reports. Seventh air victory of *Stabsfeldwebel* Josef Kiss of *Flik* 24/F, flying KD 28.37.

Due to a lack of fuel, Kiss landed just short of the Pergine airfield, badly damaging his machine (this was the last front-line deployment of KD 28.37), and it was later written off.

Oct. 25: Nieuport single-seater, crashed in the Valstagna–Monte Forcellona area. Confirmed by infantry reports, by the crew of Hansa-Brandenburg C.I 129.22 of *Flik* 21/D, as well as by the commander of the *k.u.k Luftfahrtruppen* with the *11. Armee, Rittm. FP* Wilhelm Weldin. Fourth air victory of *FP Zugsf.* Lajos Telessy of *Flik* 21/D, flying KD 28.71.

Nov. 24: SPAD single-seater, crashed south of Valstagna, confirmed by the crew of a reconnaissance airplane and by the commander of the *k.u.k Luftfahrtruppen* with the *11. Armee, Rittm. FP* Wilhelm Weldin. Fourth air victory of *FP Feldw.* Karl Cislaghi of *Flik* 21/D, flying KD 28.71.

Italian Front – Carinthia 1917

All KD victories on the Carinthian front (*k.u.k. 10. Armee*) were scored by pilots of *Flik* 16, taking off from the Seebach airfield near Vilach, also from the temporary KD airfield, Watschig airfield, in the Gail valley from July, 1917 on.

Most of the air victories were achieved with KD 28.30 which, incidentally, had a tiltable Schwarzlose 7/12 machine gun (28.26 also had such a machine-gun configuration).

The Italian airplanes defeated by the KD type were all either forced to land or crashed behind the enemy lines. The Italian casualty reports used were thoroughly checked and are hence considered — with some circumspection — reliable.

May 25: Caproni bomber of *5a Squadriglia da bombardemento,* attacked in the Kneza–Ponikve–Baza di Modreja area at approximately 1000 hours. After being given roughly 250 rounds, in the course of which the rear engine with the pusher propeller stopped, the machine gunner also collapsed motionless in the nacelle. The Caproni then made a sharp left-hand turn in order to fend off the attack with its front machine-gun.

Feldpilot Stabsfeldwebel Josef Kiss in the cockpit of KD 28.37, *Flik* 24, Summer 1917.

Right: Pergine airfield, June 1917, Emperor Karl in conversation with *Feldpilot* Josef Kiss (left). Photo: Bernhard Tötschinger.

Bottom right: *Feldpilot Feldwebel* Josef Kiss (++) in conversation with an Austro-Hungarian general on 14 June 1917 (both far right). Left: the crew of the downed Italian Farman reconnaissance airplane. In the foreground is KD 28.37. Photo: Karl Meindl.

Bottom: *Feldpilot* Josef Kiss (++ with only his back visible) celebrating his sixth air victory on 13 July 1917 with the NCO pilots of *Fliks* 21 and 24. Photo: Othmar Wolfan via Karl Meindl.

Below: Airplanes of *Fluggeschwader* I landing on the Divacca airfield; the three lower airplanes are Hansa-Brandenburg C.Is, the upper two airplanes are KD single-seaters. Photo: Josef Bangerl via Karl Meindl.

In the meantime, the KD pilot cleared a machine-gun jam that had just occurred (for this reason KD 28.30 must have already had a tiltable machine-gun by the end of May 1917!) and, during his second attack, shot the rest of his machine-gun belt into the bomber which was already going down. The pilots, first pilot *Tenente pilota* Ettore Della Cella, and second pilot *Sergente pilota* Vittorio Romagnino, still succeeded in making an emergency landing behind the Italian lines. (The airplane, badly shot up, was dismantled and salvaged.) The machine gunner, Emilio Blesso, an experienced air fighter, was killed in combat.

The commander of the *k.u.k. XV. Korps* (*Feldmarschalleutnant* Karl Schotti) observed the entire course of the air battle and enthusiastically confirmed the claim (enemy forced to land in the area of Globocak–Jeza–Monte Hum). First air victory of *FP Feldw.* Erwin Funk (28.30). For this, the pilot received the Gold Bravery Medal. On this particular day, *Feldwebel* Funk was to fly KD 28.30 to *Flik* 34 in the Wippach

Left, above and below left, and above: Photos of the Savoia Pomilio S.P.2 reconnaissance plane, No. 4530, shot down by *Feldpilot* Josef Kiss, 13 July 1917, Levico, Val Sugana. All Photos: Josef Bangerl via Karl Meindl.

Former KD pilots as flight instructors at the Campoformido field aviation school, Summer 1918. From left to right: Josef Siegel, Vinzenz Magerl, Alexander Markus, Hermann Richter. Photo: J. Arigi.

Generaloberst Svetozar Boroevic von Bojna, commanding officer of the *Isonzo Armee,* decorating NCO pilots of *Flik* 41/J. In the middle of the photo is *Hauptmann* Brumowski. Photo: Dr. Peter Schiemer.

The NCO pilots of *Flik* 41/J: *Feldpilot* Karl Kaszala (with white trousers), 8 victories; pilot at right is *Feldpilot Feldwebel* Radames Iskra (POW 6 November 1917); pilot at left is *Feldpilot Feldwebel* Josef Novák. Photo: Dr. Peter Schiemer.

Valley (Isonzo front) to strengthen this unit with an additional single-seat fighter. With a full load of ammunition, he took off for his transport flight and thereby encountered the Caproni (this comes from the pilot's combat flight log).

July 14: Farman reconnaissance airplane, crashed in flames over Monte Crete–Monte Cullar, at approximately 1130 hours; confirmed by the *k.u.k. 29. Gebirgs-Brigade* (Mountain Brigade) as the seventh air victory of *FP Hptm.* Raoul Stojsavljevic, commander of *Flik* 16, flying KD 28.30.

July 23: Farman reconnaissance airplane of *29a Squadriglia da ricognizione,* attacked during air combat at approximately 1045 hours over Trögl (Plöcken Pass) and forced into a dive behind the Italian lines. Confirmed as the eighth air victory of *FP Hptm.* Raoul Stojsavljevic, commander of *Flik* 16, flying KD 28.30.

According to Italian accounts, the crew members were

23 August 1917: The captured Italian fliers (+ *Sergente pilota* Malaspina, ++ *Tenente osservatore* Mazzarino, 45a *Squadriglia*) Sesana airfield. The Austrian pilot with the white trousers is Linke-Crawford. Photo: Bernhard Tötschinger.

Hauptmann Brumowski (at right), the Archduchess Maria Theresia (middle) and *Oberleutnant* Benno Fiala *Ritter* von Fernbrugg, 26 July 1917, Sesana airfield. Photo: War Archive, Vienna.

wounded but were still able to land at their own airfield. Crew members were *Sergente pilota* Ettore Scuri and *Tenente osservatore* Pietro Sinigaglia.

The air victories of 14 and 23 July 1917, along with his KD photo single-seater flights, are mentioned as grounds for awarding the Knight's Cross of the Order of Leopold with War Decoration and Swords to *FP Hptm.* Raoul Stojsavljevic.

Sep. 7: SAML reconnaissance airplane of *113a Squadriglia da ricognizione,* forced to land near Cleulis (in the But Valley). Confirmed by the *k.u.k. 49. Infanterie-Division* as the ninth air victory of *FP Hptm.* Raoul Stojsavljevic, commander of *Flik* 16, flying a KD. According to Italian sources, the SAML two-seater landed near Koderkohe. Crew member *Capitano pilota* (*Tenente,* according to other accounts) Innocenzo Paroli was uninjured; *Tenente osservatore* Livio Palazzo was killed. Airplane heavily damaged.

Sep. 23: Nieuport single-seater, forced to land in the Fella valley. There is no report concerning this in the report of *Flik* 16 for September 1917; only the airplane brought down by *Hptm.* Stojsavljevic on 7 September 1917 is cited. The sole source mentioning this air victory is the fascicle, *Luftsiege,* in the Austrian War Archives (*Oberst a.D.* Erich Kahlen, author). In the case of later confirmation by front-line reports, this would have been the fourth air victory of *FP Offstv.* Josef Siegel of *Flik* 16, flying an unknown KD.

Concerning the air victories of *Flik* 16 on 7 and 23 September 1917: In September 1917, *Flik* 16 still did not have any Albatros D.III single-seat fighters; this type was first allocated in November 1917. On the other hand, that same month, four KDs of Series 28 at *Flik* 16 were identified:

28.20—used only as a trainer
28.26—combat airplane
28.30—combat airplane
28.58—combat airplane

Thus, only 28.26, 28.30 or 28.58 could be considered for the victories cited.

The subject of numbering the air victories by the Austro-Hungarian pilots is a problematic one. The aircraft downed by Stojsavljevic on 7 September 1917 is listed as his fifth confirmed air victory by the *k.u.k. 10. Armeekommando;* apparently his earlier victories scored while a two-seater pilot had been overlooked.

Romanian Front (NE Transylvania) - 1917

All air victories with the KD were scored by pilots of *Jagdflieger Detachment Hauptmann* Nikitsch. In addition to the confirmations detailed, all air victories were further confirmed by the *Armeegruppe General der Infanterie von Gerok.*

Aug. 9: Nieuport single-seater fighter, forced to land in enemy territory near Grosesci at approximately 1800 hours. Confirmed as third air victory of *FP Hptm.* Karl Nikitsch, flying KD 28.29, by the *Artilleriegruppe Oberst Holländer,* and by observer *Lt.i.d.Res.* Fritz Pisko of *Flik* 44/F, in Hansa-Brandenburg C. I 69.33. (KD 28.29 with synchronized machine gun, Bernatzik synchronization.)

Aug. 11: Nieuport single-seater fighter, crashed in enemy territory near Onesti at approximately 1130 hours, cartwheeling several times. Confirmed by the *Stabsoffizier der k.u.k. Luftfahrtruppen* with the *k.u.k. 1. Armeekommando, Oberstleutnant* Georg Prinz zu Schwarzenberg, as fourth air victory of *FP Hptm.* Nikitsch (KD 28.29) and first air victory of *FP Zugsf.* Johann Hramann, flying Albatros D.III 53.24.

Divacca airfield, *Fluggeschwader* I, March 1917. Photo: Josef Bangerl via Karl Meindl.

A KD (Series 65.7), Seebach airfield, Spring 1917, *Flik* 16. Photo: War Archive, Vienna.

Seebach airfield, *Flik* 16, with KD (left) and two Hansa-Brandenburg C.Is, Summer 1917. Photo: War Archive, Vienna.

Aug. 30: Farman reconnaissance airplane (two-seater) shot down behind the enemy lines near Tirgul-Ocna at approximately 1100 hours. Confirmed as fifth air victory of *Hptm.* Nikitsch (28.29) by the *k.u.k. VIII. Korpskommando,* infantry reports, as well as by the *5. Batterie* (Commander *Oblt.* Heinrich Hardung-Hardung and *Lt.* Haas) of the *k.u.k. Feldkanonenregiment Nr. 4* (Field Gun Regiment No. 4).

Remarks: The crash sites given for the enemy aircraft are situated in the northeastern section of the Romanian province of Moldau. This section of the Romanian front was held by the Russian Ninth Army.

Russian as well as Romanian aviation units were flying in the combat zone. These units had been allocated single-seater fighters, which were being flown predominately by well-trained French and Russian fighter pilots. Detailed accounts as to the types of enemy aircraft shot down cannot be made, since the equipment of the Russian and Romanian units was not uniform. Up to four or five different types were being used in one squadron, from antiquated "crates" to the most modern equipment. The single-seater fighters shot down certainly met all necessary requirements and were at least equal to, if not superior to, the KD.

Incidentally, KD 28.47 was also used at the *Jagdflieger Detachment;* presumably it too had Bernatzik synchronization (at least the known photos of KD 28.47 do not show a VK II canister).

Individual Aircraft Histories

Brandenburg D.I Series 28

(with 185 HP Austro-Daimler Engine)

28.01: 8 February 1917 to *Fluggeschwader* I. On 1 March 1917, *Feldpilot Hptm.* Franz Rabitsch (CO of *Fluggeschwader* I) crashed fatally on the Divacca airfield during a test flight (see photo below). The airplane was destroyed and written off on 10 April 1917. According to *Feldpilot* Julius Arigi's account, the crash was not an accident, but rather an intentional suicide by *Hptm.* Rabitsch, who had taken the death of a flying comrade very much to heart.

28.02: 8 February 1917 to *Flep* 1; February 1917 to *Flik* 12; the personal airplane of *Feldpilot Oblt.* Frank Linke-Crawford from 24 February to 25 May 1917. On 25 May 1917, in air combat with two SPAD single-seaters, it was badly shot up and presumably sent off for repair. On 4 October 1917, following repairs at *Flep* 9 in Fiume, it went to *Flik* 42/J, and was written off in February 1918. Despite being written off, it was repaired at the AL-MA firm in Prague; in August 1918 it was found as a trainer without an engine with *Flek* 6.

28.03: 8 February 1917 to *Flik* 19, an escort fighter until the end of April 1917; to *Flik* 41/J at the beginning of May 1917. Found as a pursuit trainer with *Flek* 6 in March 1918; written off May 1918 after a landing accident.

28.04: 8 February 1917 to *Flik* 34, an escort fighter until October 1917. That same month it was sent to *Flep* 9 in Fiume as unusable for combat. Personal airplane of *Feldpilot Feldwebel* Alexander Renner.

28.05: 28 February 1917 to *Flik* 23, an escort fighter at least until 19 June 1917 (personal machine of *Feldpilot Feldwebel* Ernst Franz); later handed over to *Flik* 28. It was damaged at the beginning of November 1917 by *Korporal* Kauer while landing at the new Torresella airfield, and sent to *Flep* 9 for repairs. Subsequently a trainer with *Flek* 6, supposedly written off in March 1918. According to reports, 28.05 had its own machine gun installed at *Flik* 28.

28.06: 28 February 1917 to *Fluggeschwader* I, deployed at the front as an escort and interceptor fighter until 17 September 1917. With three air victories, it was one of the most successful KD aircraft.

28.07: 28 February to *Flik* 35, then a trainer with *Flek* 6; written off 28 April 1917 following a landing accident.

28.08: 28 February 1917 to *Fluggeschwader* I; in March 1917 it was repaired twice at *Fluggeschwader* I following accidents. From July 1917 it was with *Flik* 41/J; on 4 October 1917 it was again assigned to *Flik* 41/J, following repairs at *Flep* 9.

28.09: 24 February 1917 to *Flik* 41/J in Strasshof; deployed at the front during April 1917 with *Flik* 41/J. It was damaged that same month and was written off as of 26 May 1917.

28.10: 24 February 1917 to *Flik* 41/J in Strasshof (unit was in the process of forming); deployed at the front with *Flik* 41/J until at least 1 September 1917. First personal airplane of *Feldpilot Hptm.* Godwin Brumowski.

28.11: 24 February 1917 to *Flik* 41/J in Strasshof; determined as being deployed at the front until at least the middle or end of May 1917. Written off 12 July 1917, after a "flight error."

KD 28.01, destroyed in *Hauptmann* Rabitsch's fatal crash on 1 March 1917, Divacca airfield. Photo: Josef Bangerl via Karl Meindl.

Above and below: Crash of KD 28.14, Flik 41/J, April 1917. Both Photos: War Archive, Vienna.

KD 28.37 in flight over the Pergine airfield, Val Sugana, Summer 1917, *Flik* 24. Photo: War Archive, Vienna.

28.12: 24 February 1917 to *Flik* 41/J in Strasshof; written off 26 May 1917 after a crash due to an engine defect.
28.13: 24 February 1917 to *Flik* 41/J in Strasshof; flown by almost all of the pilots of *Flik* 41/J (see profiles). Written off 3 November 1917; nevertheless, in August 1918 it was found as a trainer without an engine at *Flek* 6.
28.14: 2 March 1917 to *Flik* 41/J. On 21 April 1917 it crashed at the Sesana airfield, breaking up completely, and *Stabsfeldwebel* Kurt Gruber was severely injured; written off May 1917. (Crash was not 21 May as given elsewhere.)
28.15: 2 March 1917 to *Flik* 41/J. On 24 April 1917, Chief Pilot *Feldpilot Oblt.* Leo Bisce was injured when he crashed while landing at the Sesana airfield. After being repaired at *Flep* 1 in Marburg, the airplane was again sent to *Flik* 41/J. On 6 August 1917, it went to *Flep* 9 in Fiume for repairs due to a bad accident and was written off by 11 September 1917. (Crashed by Linke-Crawford?)
28.16: Delivered 1917 as a pursuit trainer with *Flek* 6; from Summer 1918 it was found with *Abwehrstaffel* I in Wiener-Neustadt; was still on the rolls of the *Flet* on 31 August 1918.
28.17: 12 March 1917 to *Flik* 41/J. On 10 May 1917 *Feldpilot Zugsführer* Rudolf Stöhr was shot down in air combat by *Capitano pilota* Francesco Baracca and severely injured. The airplane was heavily damaged during his emergency landing in his own territory and was written off 12 July 1917.
28.18: 12 March 1917 to *Flik* 42/J; flown mainly by *Feldpilot Feldwebel* Vinzenz Magerl. On 31 July 1917, while on an escort flight, the carburetor caught fire over Comen. The airplane was demolished during the resulting emergency landing and *Zugsführer* Dezsö Jäger was injured. Airplane was written off 1 September 1917.
28.19: 29 May 1917 to *Flik* 41/J, written off by 31 July 1917.
28.20: 29 March 1917 to *Flik* 42/J; mentioned in September 1917 as a trainer with poor climb capability with *Flik* 16/D; written off 22 December 1917 following a landing accident. Despite being written off, it was being used as a trainer without an engine at *Flek* 6 in August 1918.
28.21: 3 April 1917 to *Flik* 42/J; written off 12 July 1917 following a landing accident.
28.22: 3 April 1917 to *Flik* 42/J. On 17 May 1917, it took a direct flak hit during an offensive patrol over Merna. With a dead engine and its undercarriage shot to pieces, it made an emergency landing behind its own lines, during which it was badly damaged. The pilot was uninjured. After repair, it was deployed for a short time with *Flik* 41/J, then written off as of 25 August 1917.
28.23: 29 March 1917 to *Flik* 42/J; 4 October 1917 to *Flik* 41/J, following repairs at *Flep* 9 in Fiume, and flown almost exclusively by *Korporal* Ignaz Pillwein during escort flights in October 1917. On 1 February 1918, it was sent to the aviation arsenal, in damaged condition and without an engine. It was written off at the Aspern supply depot on 1 October 1918.
28.24: 29 March 1917 to *Flik* 42/J. In Spring 1917, it caught fire as a result of a carburetor fire and was damaged during the resulting emergency landing; written off as of 12 July 1917. KD 28.24 was never with *Flik* 16 as stated elsewhere.

KD 28.22. Photo: Walter Schroeder.

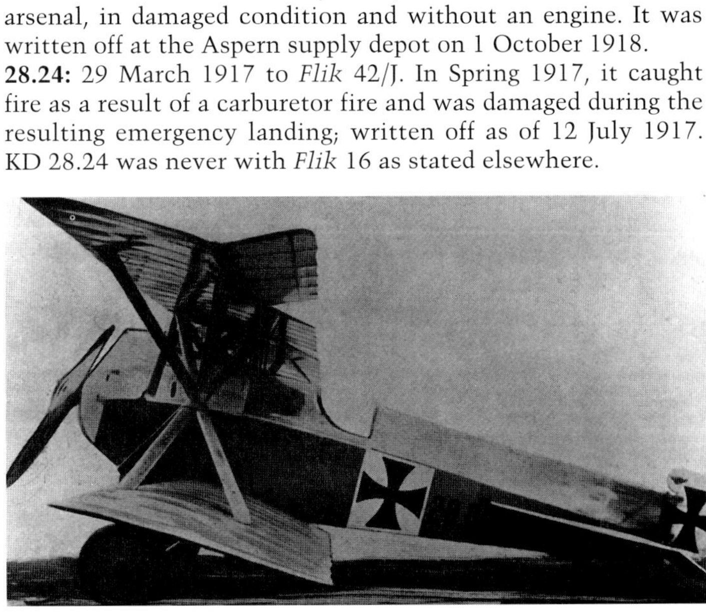

KD 28.02, *Flik* 12, Spring 1917. Photo: R. Cavigioli.

KD 28.37, *Flik* 12, Summer 1917, Haidenschaft airfield. Photo: Franz Selinger.

An Italian Farman shot down on the Isonzo front, Summer 1917. Photo: J. Arigi.

28.25: 7 April 1917 to *Flik* 42/J. On 23 May 1917, it made an emergency landing near Tomaj due to an overheated engine and the ensuing carburetor fire. The airplane burned on the ground, and *Feldpilot Lt.i.d.Res.* Richard Melzer was severely injured; airplane written off 12 July 1917.
28.26: Delivered 20 January 1917; accepted April 1917; to *Flep* 1 in Marburg on 24 April 1917. Deployed at the front as an escort fighter, interceptor aircraft and photography single-seater until mid-January 1918. In the afternoon hours of 11 January 1918, while flying a photography mission, *Feldpilot Hptm.* Raoul Stojsavljevic (CO of *Flik* 16/D) was severely wounded by phosphorus ammunition in air combat with five enemy aircraft and made an emergency landing behind his own lines near Seren. The airplane was badly damaged and written off 21 March 1918.

This was probably the aerial victory shot down in flames near Seren at 1235 hours (British time) by British pilots Lieutenant K.B. Montgomery and Second Lieutenant A.F. Lingard of 66 Squadron, flying Sopwith Camels. However, three Italian pilots also claim to have downed 28.26. According to the December 1917 report by *Flik* 16/D, 28.26 did not have camouflage paint.
28.27: 7 April 1917 to *Flik* 41/J; deployment at the front traceable until July 1917.
28.28: Delivered 20 January 1917. In July 1917, it was with *Flik* 23 as an escort fighter; with *Flik* 28 from August 1918, being flown by *Feldpilot Feldwebel* Alphons Korsitzky; later handed over to *Flik* 46/F.
28.29: Delivered 31 January 1917. Delivered to the *k.u.k. 1. Armee* as of 12 July 1917; deployed at the front with *Fliegerdetachment Hauptmann Nikitsch* as an escort and interceptor plane. Upon leaving for the Italian front, Nikitsch took 28.29 along and used it for numerous flights until 24 November 1917. During a practice flight, *Feldpilot Hptm.* Karl Nikitsch (at that time CO of *Flik* 63/J) crashed on the Cinto Caomaggiore airfield from an altitude of 20 meters during which he was badly injured and the aircraft was demolished. Written off 21 March 1918.

Armament: 28.29 had a Bernatzik synchronization system on its Schwarzlose M7/12.
28.30: Delivered 17 February 1917; accepted by *Flik* 16 April 1917; used as an escort fighter and photography single-seater until January 1918.

Armament: an over-wing machine gun that could be pulled down to load.
28.31: Delivered 17 February 1917; to *Flep* 1 in Marburg on14 June 1917; utilized at *Flik* 28/D as an escort fighter. In October 1917 (exact date unknown) after an escort flight, it was involved in a landing accident in which *Korporal* Max Kauer was uninjured. Following repairs at *Flep* 9, it was used as a trainer with *Flek* 9, and written off 3 March 1918.
28.32: Delivered 17 February 1917; deployed with *Flik* 42/J. In July 1917, the undercarriage failed on take-off and the airplane was demolished upon landing. Written off 1 September 1917.
28.33: Delivered 21 February 1917; deployed with *Flik* 41/J. Following an accident at *Flep* 9 in Fiume, it was written off on 25 August 1917.
28.34: Delivered 17 February; accepted May 1917; first deployed with *Flik* 42/J, then an escort fighter with *Flik* 2/D in Autumn 1917. In June 1918, it was a trainer without an engine with *Flek* 6.

Armament: according to the October 1917 report of *Flik* 2/D, 28.34 had an individual machine gun assembly which was constructed by the *Flik*, a Schwarzlose M7/12 machine gun which fired over the arc of the propeller.
28.35: Delivered 21 February 1917; accepted May 1917; deployed with *Flik* 21 on the South Tyrolean front. On 14 June 1917 (Battle of Ortigara) *Feldpilot Feldwebel* Richard Schuster was badly wounded in air combat and the airplane was demolished in the course of his emergency landing in the vicinity of Vezzena. Written off July 1917.

This was probably an aerial victory by the Italian fighter ace Nicelli, who scored his first air victory on this day. Schuster was attacked by four Italian Nieuports.
28.36: Delivered 17 February 1917; accepted May 1917; deployed from June 1917 with *Fluggeschwader* I. Later it was in the rear for the purpose of repairs, going to *Flep* 3 in Trient on 13 November 1917.
28.37: Delivered 21 February 1917; accepted May 1917; deployed at the front from June 1917 with *Flik* 24, the personal airplane of *Feldpilot Stabsfeldwebel* Josef Kiss. After a combat patrol on 14 September 1917, due to a lack of fuel, it made an emergency landing outside the Pergine airfield in which it was damaged. It was then sent to Aspern on 28 September 1917, without an engine, and written off on 27 October 1917. With four victories, 28.37 was the most successful KD fighter on the Tyrolean front.

Armament: 28.37 had a Bernatzik synchronization system.
28.38: Delivered 26 February 1917. On 8 June 1917, it came to *Flik* 12 in rundown condition where it was test-flown by *Feldpilot Oblt.* Linke-Crawford and used on numerous missions. Later the personal airplane of *Oblt.i.d.Res.* Benno Fiala *Ritter* von Fernbrugg (5 air victories in this aircraft). Damaged in November 1917; following repairs, in August 1918, it was a trainer without an engine at *Flek* 6.
28.39: Delivered 26 February 1917; went to *Flik* 41/J at the end of May 1917; written off 25 August 1917.
28.40: Delivered 28 February 1917. At the end of May 1917 it

Offstv. Kiss in 28.37 at Pergine, 14 September 1917. Photo: Franz Selinger Collection via Walter Schroeder.

Crash of *FP Korporal* Aladar Balász of Flik 32 in 28.51 on the St.Veit airfield near Wippach. Photo: Walter Schroeder.

KD 28.43, *Flik* 35, *Feldpilot Feldwebel* Karl Sailer, shot down in flames and KIA on 14 July 1917. Photo: War Archive, Vienna.

went to *Flik* 41/J where it was flown by almost all of the pilots of *Flik* 41/J. In August 1918, it was at *Flek* 6 being used as a trainer without an engine; written off October 1918.

28.41: Mid-May 1917 to *Flik* 41/J; 1 October 1918, found at the Aspern supply depot without an engine.

28.42: Accepted May 1917; deployed with *Flik* 42/J. On 10 July 1917, *Zugsführer* Dezsö Jäger forced a Caudron to go down in a dive behind the Italian lines (unconfirmed). In the evening hours of that same day, *Feldpilot Zugsführer* Ferdinand Udvardy shot down a SPAD VII and a Nieuport fighter over Monte Sabotino. In July 1917, it was damaged during landing; after repairs at Flep 9, it was once again allocated to *Flik* 42/J on 4 October 1917. August 1918, found as a trainer without an engine at *Flek* 6.

28.43: Delivered 1917; with *Flik* 35 as an escort fighter from June 1917. On 14 July 1917, *Feldpilot Feldwebel* Karl Sailer was shot down in flames and killed in air combat near Gorjansko. The airplane burned on the ground, and was written off 1 September 1917.

This was an air victory of the Italian fighter ace *Tenente pilota* Fulco Ruffo di Calabria of 91a *Squadriglia da caccia.*

28.44: Delivered 1917. From the end of June 1917, it was flown by almost all of the pilots of *Flik* 41/J. On 4 October 1917, following repairs at *Flep* 9, it was again allocated to *Flik* 41/J. During a test flight on 16 December 1917, it was involved in a fatal crash on the Torresella airfield, as a result of total wing failure. The airplane was totally demolished and was written off March 1918. An investigation revealed indisputable sabotage; the spars of the wings had first been sawed into and then carefully glued.

28.45: Accepted May 1917; deployed with *Flik* 42/J; following a landing accident, it was written off 28 July 1917.

28.46: Accepted 28 August 1917; deployed with *Flik* 101/G as an escort fighter until January 1918.

Armament: 28.46 had a tiltable machine gun for armament.

28.47: Mid-July 1917, sent to the Transylvanian front, to *Fliegerdetachment Hauptmann Nikitsch,* being utilized as an escort and interceptor fighter aircraft. On 21 August 1917, following an escort flight for a reconnaissance airplane of *Flik* 29, it crashed on the Kézdi-Vásárhely airfield during landing. *Feldpilot Zugsführer* Johann Hramann was badly injured in the crash, and the airplane was demolished. Written off 4 October 1917.

28.48: Converted to test airplane 20.14 (see pages 73–76 of *Austro-Hungarian Army Aircraft of World War One*); damaged on 16 January 1917 during a test flight by *Hptm.* Nikitsch at the aviation arsenal.

28.49: In July 1917, this airplane came to the Acceptance Department of the aviation arsenal for testing of fuselage failure (having been previously damaged, it was therefore no longer considered for flying purposes). Shown on the rolls of the *Flet* on 31 August 1918.

28.50: Renumbered 20.15. See *Austro-Hungarian Army Aircraft of World War One,* page 76.

28.51: Accepted May 1917. From June 1917, it served as an escort airplane with *Flik* 32. It was demolished in August 1917 as a result of a Bora (see glossary) while landing at the airfield. *Feldpilot Korporal* Aladar Balasz was uninjured; the airplane was written off 2 October 1917.

28.52: Accepted May 1917; at *Flik* 41/J from mid-August 1917. On 21 August 1918, it crashed on the Sesana airfield as a result of a carburetor fire, after taking off on a combat patrol. Both the airplane and *Feldpilot Zugsführer* Ferdinand Jaschek were consumed by the fire on the ground. Written off 4 October 1917.

28.53: Accepted May 1917; sent to *Flep* 1 in Marburg in mid-June 1917; deployed with *Fliks* 34 and 41/J; written off March 1918.

28.54: Accepted May 1917; deployed from July 1917 with *Flik* 46/F as an escort fighter; damaged at the end of July 1917 during landing and sent to *Flep* 9 in Fiume for repair. Deployed from the end of September 1917 to 28 January 1918 with *Fluggeschwader* I (Flik 101/G) as an escort fighter (flown mainly by *Stabsfeldwebel* Jaroslav Samek). On 29 July 1918, *Zugsführer* Wandl of *Flek* 6 crashed during a practice flight near Innerfahrafeld in the Traisen Valley. The pilot was injured; the airplane was heavily damaged and written off September 1918.

28.55: Delivered 1917; in mid-June 1917, it went to *Fluggeschwader* I as an escort fighter. On 26 July 1917, during

The grave of pilot *Korporal* Altmann Herzog, *Flik* 41/J, killed 16 December 1917. Photo: Bernhard Tötschinger.

The grave of *Feldpilot Zugsführer* Ferdinand Jaschek, *Flik* 41/J, killed 21 August 1917. Photo: War Archive, Vienna.

a flight to protect a radio aircraft, it was badly shot up in air combat with a SPAD VII; *Feldpilot Feldwebel* Karl Dovolil was uninjured. Heavily damaged during a take-off in September 1917, it was written off 31 October 1917.

28.56: Delivered 1917; went to *Flik* 42/J in mid-June 1917; in March 1918, a trainer with *Flek* 6.

28.57: Delivered 1917; went to *Flik* 41/J in mid-June 1917, where it was flown by almost all the pilots of the unit. After repairs at *Flep* 9, on 4 October 1917, it was again allocated to *Flik* 41/J.

28.58: Delivered 1917; went to *Flik* 42/J at the end of June 1917; then to *Flik* 41/J and *Flik* 16. In September 1917, it was sent, damaged, to *Flep* 1. Following repair, it was used as a trainer with *Flek* 5, and written off in September 1918.

28.59: Delivered 1917; in mid-June 1917, it went to *Flep* 1 in Marburg; August 1918, it was with *Flik* 28 as an escort fighter. On 19 August 1917, it made an emergency landing near Rubbia on account of a shot-up gas tank after downing a SPAD VII. *Feldpilot Feldwebel* Alphons Korsitzky was uninjured but the airplane was badly damaged, and was written off 4 October 1917. *Flik* 28 placed this KD at the disposal of a Wippach company as a loaner airplane (unit unknown.)

28.60: Delivered 1917. As of 25 September 1917, it went first to the Isonzo front and was briefly with a *Flik* of the *2. Isonzo Armee*. In 1918, it was repaired at the AL-MA Firm in Prague. In August 1918, it was a trainer without an engine at *Flek* 6.

28.61: Delivered 1917; it went to the Isonzo front on 30 June 1917; in September 1918, it was with *Flik* 28/D. Following an escort flight on 15 September 1917, it overturned on the airfield during landing and was heavily damaged; it was then written off on 31 October 1917. *Feldpilot Feldwebel* Alphons Korsitzky was uninjured in the crash.

28.62: Delivered 1917; in mid-July 1917, it went to the Isonzo front and was an escort fighter with *Flik* 46/F; the personal airplane of Chief Pilot *Oblt.i.d.Res.* Dr. Josef Hoffmann *Ritter* von Ostenhof. On 26 October 1917, it was sent, damaged, to *Flep* 9; written off 3 March 1918.

28.63: Delivered 1917; to *Flep* 3 in Trient on 12 July 1917. From at least 12 October to 6 November 1917, it was with *Flik* 17/D as an escort fighter. During the same month, it was heavily damaged during an emergency landing, and was written off in December 1917. At *Flik* 17/D, it was flown by *Feldpilot Lt.* Max Schoßleitner.

28.64: Delivered 1917; on 10 July 1917, it went to *Flep* 1. In September 1917, it was at *Flik* 28/D as an escort fighter. After being repaired at *Flep* 9 in Fiume, it was allocated to *Flik* 42/J on 4 October 1917; bore a swastika cross insignia.

28.65: Delivered 1917; went to *Flik* 41/J on 10 July 1917, where it was flown by almost all of the pilots in July, August, and through mid-September 1917.

28.66: Delivered 1917; at the end of July 1917, it went to *Flik* 12 and was test-flown by *Feldpilot Oblt.* Linke-Crawford. On 14 August 1917, during a defensive patrol, the radiator was shot up in air combat with three Caproni bombers. Despite an extremely strong Bora wind, it made an emergency landing at its own airfield with a dead engine. On the ground, the airplane was caught by the Bora wind and demolished; *Feldpilot Oblt.i.d.Res.* Benno Fiala *Ritter* von Fernbrugg was uninjured. Airplane written off 3 November 1917.

28.67: Delivered 1917; on 4 August 1917, it went to *Flik* 36 as an escort fighter. Later it was used a trainer at *Flek* 6. On 19 August 1918, *Korporal* Knirsch crashed fatally during a practice flight as a result of the airplane stalling in a turn. Totally demolished, it was written off in October 1918.

28.68: Delivered 1917; on 31 July 1917, sent to *Flik* 36 (German *Donau Armee*, Romania); written off March 1918. The fuselage section of 28.68 is displayed in the Museum of Technology in Prague.

Above: Visit of the Archduchess Maria Theresa to *Flik* 41/J at Sesana on 26 July 1917. Photo: Walter Schroeder.

28.69: Delivered 1917; on 10 July 1917, it went to the Isonzo front. Deployed from the beginning of August 1917 with *Flik* 41/J; one of the many personal airplanes of *Feldpilot Hptm.* Godwin Brumowski, who scored seven air victories with this airplane. In September 1917, it was heavily damaged in a landing accident and was written off 3 November 1917.

28.70: Delivered 1917; on 31 July 1917, it went to the Romanian front.

28.71: Delivered 1917; went to the South Tyrol on12 July 1917; deployed in the autumn of 1917 with *Flik* 21/D. On 14 January 1918, in air combat with two SPADs over Monte Melago, *Feldpilot Feldwebel* Karl Cislaghi was shot down between the lines and made an emergency landing, after which he was taken prisoner by the Italians. Since the airplane could not be brought in due to the difficult terrain, it was blown up by a patrol of the *k.u.k. Infanterie-Regiment Nr.27* during the night of 15 January 1918. Probably an aerial victory of the Italian fighter ace *Tenente pilota* Silvio Scaroni.

28.72: Delivered 1917; used as a trainer at *Flek* 6; found with *Abwehrstaffel* I (defense flying unit) in Wiener-Neustadt in the summer of 1918.

28.73: Renumbered 20.16. See *Austro-Hungarian Army Aircraft of World War One,* pages 76–78.

Godwin Brumowski with a camouflaged Phönix-built D.I, probably his favorite, 28.69, *Flik* 41/J, Sesana 1917. Brumowski became the leading Austro-Hungarian ace with 35 confirmed air victories. This is the only known photo clearly showing a camouflaged Brandenburg D.I. Photo: M. O'Connor.

Above: Brandenburg D.I(Ph) 28.01? with pilot Edmund Sparmann at Aspern.

Right: Brandenburg D.I(Ph) 28.01.

Below: Brandenburg D.I(Ph) 28.02.

Above and right: Brandenburg D.I(Ph) 28.02.

Above: Brandenburg D.I(Ph) 28.04.

Left: Although the serial number on this aircraft appears like it could be 28.01, the shape of the tail suggests strongly that it is actually Brandenburg D.I(Ph) 28.04 shown above.

Brandenburg D.I(Ph) 28.06 in flight.

Above: Brandenburg D.I(Ph) 28.10 with original tail surfaces. See pages 48 and 52 for additional views.

Right: Brandenburg D.I(Ph) 28.15 crashed by Frank Linke-Crawford at *Flik* 41/J, summer 1917. See page 62 for another view of this accident. See pages 48 and 52 for views of this aircraft with its original tail surfaces.

Below: Brandenburg D.I(Ph) 28.13 on its nose. This view shows that the machine gun in the VK II canister was off-set slightly to the right of the centerline.

Above: Brandenburg D.I(Ph) of *Flik* 36.

Above: Brandenburg D.I(Ph).

Oblt. Urban with Brandenburg D.I(Ph) 28.24. This aircraft has a transparent windshield and a curved rail to lower the gun to clear jams.

Oblt. Urban on 14 August 1917 in Brandenburg D.I(Ph) 28.24. There is an additional metal strap for the rear machine-gun support compared with the photo above.

Above: Brandenburg D.I(Ph) 28.28.

Left: Brandenburg D.I(Ph) 28.26 as a *fotoeinsitzer* with *Flik* 16, December 1917.

Below: Brandenburg D.I(Ph) 28.30 and *Hptm.* Raoul Stojsavljevic, CO of *Flik* 16. Stojsavljevic pioneered the use of single-seat aircraft for high-speed photo-reconnaissance in this aircraft.

Brandenburg D.I(Ph) 28.30 and *Hptm.* Raoul Stojsavljevic, *Flik* 16, on 14 August 1917. There is a transparent windshield and a curved rail on which the machine gun could be lowered to clear jams, an impossibility with the standard VK II gun canister. The black streak is an oil smear from the machine gun.

Below: Brandenburg D.I(Ph).

Brandenburg D.I(Ph) 28.33.

Below and bottom: Brandenburg D.I(Ph) 28.37.

Above and above right: Brandenburg D.I(Ph) 28.37 flown by Josef Kiss of *Flik* 24, with 19 confirmed victories the fifth-ranking ace of the Austro-Hungarian Empire, summer 1917.

Right: Brandenburg D.I(Ph) 28.37 being righted after a bad landing.

Below right: Brandenburg D.I(Ph) 28.37 right side up after a bad landing.

Below: Brandenburg D.I(Ph) 28.37.

Above and top: Brandenburg D.I(Ph).

Above left: Brandenburg D.I(Ph) 28.38 and *Oblt.* Benno Fiala *Ritter* von Fernbrugg, *Flik* 12D, autumn 1917. With 28 confirmed victories, Fiala was the third ranking ace of the Austro-Hungarian Empire.

Left: Brandenburg D.I(Ph).

Below: Brandenburg D.I(Ph) at Villach with Brandenburg C.Is.

Above: Brandenburg D.I(Ph) 28.47.

Right: Brandenburg D.I(Ph) 28.47.

Above and above right: Brandenburg D.I(Ph) 28.58. Different lighting gives very different impressions of the aircraft's color scheme.

Brandenburg D.I(Ph) 28.58.

Below: Lineup of Brandenburg D.I(Ph) fighters of *Flik* 41/J during the visit of Archduchess Maria Theresia on 26 July 1917. An Albatros D.III is at the extreme left.

Top: Brandenburg D.I(Ph).
Above: Crash of Brandenburg D.I(Ph) 28.01 on 1 March 1917.
Right: Brandenburg D.I(Ph) 28.58 in flight from another airplane.
Below right: Brandenburg D.I(Ph) 28.67.
Below: Brandenburg D.I(Ph) "6" in flight.

Lineup at *Flik* 41/J. Brandenburg D.I(Ph) 28.15 is on the left and Brandenburg D.I(Ph) 28.10 is in the center. These aircraft have the original tail surfaces.

Above: Brandenburg D.I(Ph).

Brandenburg D.I(Ph) 28.60.

A Brandenburg D.I(Ph).

Below right: A Brandenburg D.I(Ph).

Above: Brandenburg D.I(Ph) 28.13. Another view of this accident is on page 38.

Above right: Brandenburg D.I(Ph) 28.15. Other views of this accident are on pages 37 and 62.

A Brandenburg D.I(Ph).

Hptm. Raoul Stojsavljevic in Brandenburg D.I(Ph) 28.68 of *Flik* 34, February 1917. The square patch of lighter wood beneath the cockpit marks where a cross patee and its white field were scraped off to avoid providing an aiming point. Stojsavljevic, "The Iron Stoj," became an ace in this aircraft and went on to score 10 confirmed victories. The fuselage of this aircraft is preserved in the Prague Technical Museum.

Below: The crash of Brandenburg D.I(Ph) 28.05. The fin and rudder are camouflaged.

Above: Brandenburg D.I(Ph) 28.65.
Below: Probably another view of Brandenburg D.I(Ph) 28.65.

Above: The crash of Brandenburg D.I(Ph) 28.44 at *Flik* 41/J on 16 December 1917 due to sabotage. There is a white border around the wing cross but none around the rudder cross.

Brandenburg D.I(Ph) 28.62 in flight. At this time the insignia did not have a white background.

Brandenburg D.I Series 65.5
(with 160 HP Austro-Daimler Engine)

65.50: Arrived at *Flik* 4 on 17 January 1917; with this unit until February 1917; handed over to *Flik* 12 in March 1917. Presumably at the end of May or June 1917, it was severely damaged upon landing at the Wippach airfield when it collided with Hansa-Brandenburg C.I 29.5 during its landing run. (See photos.) Written off 10 July 1917.

65.51: Arrived at *Flik* 2 on 25 November 1916, where it was merely deposited, and then handed over to *Flik* 19 at the beginning of December 1916; in March 1917 to *Fluggeschwader* I. Following an accident during take-off in April 1917, it was written off 15 May 1917.

Armament: A Spandau LMG 08/15 to start with; from February 1917, a Schwarzlose M.16 in a VK II canister over the wing.

65.52: To *Flik* 4 on 25 November 1916; damaged on 9 December 1916 during a test flight by the chief pilot of *Flik* 4, *Feldpilot Oberleutnant* Ladislaus Háry; repaired at *Flik* 4. By March 1917 it was no longer at the front; it was then used as a pursuit trainer at *Flek* 6. Determined as still on the rolls of the *Flet* 31 as of 31 August 1918.

Armament while at *Flik* 4: A Schwarzlose M7/12 in a VK II canister.

65.53: Originally to be delivered to *Flik* 19; however, on 2 November 1916, it went to *Flik* 12, where it was test flown in mid-November 1916, probably by *Feldpilot Oberleutnant* Brumowski. On 26 January 1917, during a test flight by *Feldpilot Oberleutnant* Linke-Crawford, it was heavily damaged upon landing (cause was a sudden gust of wind). It was written off 20 March 1917.

Armament: One Schwarzlose M7/12 in a VK II canister.

65.54: To *Flik* 12 on 25 November 1916. Due to a missing propeller, it was first test flown in mid-January 1917. During the test flight, it was damaged and shipped off for repairs.

Armament: One Schwarzlose M7/12 in a VK II canister.

65.55: To *Flik* 34 on 29 November 1916. At the beginning of December 1916, it was damaged at *Flik* 34, however, it was repaired at *Flik* 34, and determined as still being with *Flik* 34 in February 1917. From March 1917, it was no longer at the front; later being used as a pursuit trainer at *Flek* 6; determined as still on the rolls of the *Flet* on 31 August 1918.

65.56: Intended for *Flik* 28, it was, however, damaged at *Flik* 23 in mid-November 1916, and sent to Aspern for repairs at the end of November 1916. This airplane had probably been severely damaged during a test flight by *Feldpilot Oberst* Emil Uzelac some time prior to 20 November 1916. Uzelac was badly injured (concussion, etc.) in the crash. Uzelac spent from 31 October to 26 November 1916 (including travel days) on the Isonzo front, on account of many complaints about the KD model.

65.57: Delivered September 1916. On 26 September 1916, German civilian pilot Heinrich Bill suffered a fatal crash in Aspern as a result of the left stabilizer surface opening up (cause was shoddy workmanship at the factory). Totally demolished, it was written off 14 December 1916.

65.58: To *Flik* 19 on 2 November 1916; with *Flik* 12 from mid-November 1916; with *Flik* 4 from the end of November 1916. Damaged at the beginning of December 1916 (crashed during a test flight), it was sent to Aspern on 12 December 1916, and written off 12 January 1917.

65.59: To *Flik* 19 on 2 November 1916 and was with this unit until February 1917.

Armament: November 1916–January 1917: One Spandau LMG 08/15; February 1917: two Spandau LMG 08/15 in a VK II canister.

Above: Two views of KD 65.50 of *Flik* 4 after colliding with Hansa-Brandenburg C.I 29.5 while landing, Wippach airfield, spring 1917. Photo: Koloman Hajos via Karl Meindl.

Heinrich Bill, German civilian pilot killed in the crash of 65.57 on 26 September 1917. Photo: Walter Schroeder.

Above: KD lineup; 28.10 in the center, 28.15 at left. Photo: Walter Schroeder.

Left: A non-commissioned pilot and mechanics of *Flik* 23 posing in front of KD 65.63 on the Prosecco airfield. The cut-back engine cowling provided better cooling in the summer. Photo: Walter Schroeder.

Below: The wreck of KD 65.61. Photo: Walter Schroeder.

65.60: From 1 December 1916 to the end of January 1917, it was used as a pursuit trainer with the newly-established *Fluggeschwader* I in Straßhof; later used as a trainer with *Flek* 6.
65.61: To *Flik* 28 on 2 November 1916. Already damaged by mid-November, it was written off on 2 January 1917.
65.62: To *Fluggeschwader* I in Straßhof on 1 December 1916, then used as a pursuit trainer with *Flek* 6. On 1 May 1918, *Oblt.* Ernst Mitter crashed fatally during a practice flight in Wiener-Neustadt as a result of a carburetor fire. The airplane burned on the ground and was written off on 20 July 1918.
65.63: Arrived at *Flik* 23 on 18 January 1917; flew escort flights during February 1917; with *Fluggeschwader* I from April 1917. In mid-May 1917 it was damaged at *Flep* 1 in Marburg; with *Flik* 16 in July 1917, and disposed of after being damaged (date unknown). At the beginning of October 1917, it was sent to the rear.
65.64: To *Flik* 28 on 18 January 1917, February 1917 saw it used as an escort fighter. From April 1917 to 20 May 1917, it was found at *Flik* 23 being used as an escort fighter. Sent to the rear as of 6 October 1917.
65.65: Delivered to *Flep* 1 in Marburg on 12 January 1917; during March–April 1917 it was with *Flik* 35; and with *Flik* 41/J from the beginning of May 1917.
65.66: Delivered to *Flep* 1 on 12 January 1917, its utilization at the front cannot be ascertained. Later used as a pursuit trainer at *Flek* 6.
65.67: To *Flik* 28 on 2 November 1916; end of December 1916, it was sent to Aspern, heavily damaged (without engine). Later a pursuit trainer at *Flek* 6.

Armament: Mid-November 1916, one Schwarzlose M7/12; end of November 1916, one Schwarzlose M.16; from mid-December 1916, once again equipped with a Schwarzlose M7/12.

65.68: To *Flik* 34 on 17 January 1917; with *Flik* 16 from the beginning of May to mid-May 1917; sent to the rear by the *Flep* in Krainerend as of 6 October 1917.
65.69: Delivered 1916; with *Flek* 6 as a pursuit trainer; determined to be on the rolls of the *Flet* on 31 August 1918, although it was supposedly written off following a landing accident in April 1918.

The four photos above show KD 65.56. The two lower photos show KD 65.56 after *Oberst* Uzelac crashed it 7 November 1916 at *Flik* 19.

Above: Godwin Brumowski in KD 65.53.

Below left and above: KD 65.53.
Below: KD 65.67.

Above and right: KD 65.59.

KD 65.60.

KD 65.60.

Left and above: KD 65.60.

Top, above, above right: KD 65.5 series.

KD 65.66.

Brandenburg D.I Series 65.70
(with 150 HP Austro-Daimler Engine)

65.70: Delivered end of 1916, determined as being a pursuit trainer with *Fleks* 6 and 19.
65.71: Delivered end of 1916; used as a fighter trainer with *Flek* 6; written off in September 1917, following a landing accident.
65.72: Accepted 20 November 1916, as a pursuit trainer with *Flek* 6; written off 10 May 1917, following a take-off accident in April 1917.
65.73: To *Flik* 2 at the end of January 1917. In February 1917 it was sent, damaged, to *Flep* 1 at Marburg. In June/July 1917, it was stored at the *Flugzeugwerk Aspern* (Aspern Aircraft Works) in damaged condition, without an engine. Following repairs, it was used as a pursuit trainer with *Flek* 6, and written off following a landing accident January 1918.
65.74: To *Flik* 4 on 16 January 1917; disposed of in February 1917 as unusable for combat; then a pursuit trainer with *Flek* 6 after repairs. On 14 February 1918, *Korporal* Koller crashed fatally during a practice flight in Wiener-Neustadt; the airplane was totally demolished and written off April 1918. Despite being written off, it was on the rolls of the *Flet* on 31 August 1918.
65.75: To *Flik* 23 on 16 January 1917. It was damaged on 16 February 1918 when it crashed from an altitude of 3000 meters in the vicinity of the Prosecco airfield, without enemy intervention. *Lt.i.d.Res.* Mirko Vrbancic was uninjured.
Armament: One VK II canister on the upper wing.
65.76: To *Flik* 28 on 16 January 1917 where it remained until April 1917. Following an unknown accident, in May 1917 it was at *Flep* 1 for the purpose of repairs; July 1917 at *Flik* 16, later a pursuit trainer at *Flek* 6.
Armament: In July 1917 at *Flik* 16, a captured Italian Revelli M.14 was installed on a trial basis.
65.77: Pursuit trainer with *Flek* 6, written off following a landing accident June 1917
65.78: To *Flik* 34 on 16 January 1917. On 31 January 1917, while being test-flown, it crashed due to engine failure; totally demolished, it was written off on 15 March 1917. *Feldpilot Oblt.* Gustav Rubritius was slightly injured in the crash.
65.79: To *Flik* 35 on 16 January 1917; mid-May 1917, it was at *Flep* 1 in Marburg for the purpose of repairs; on 6 October 1917 it was sent to the rear from *Flik* 23. It was then used as a training plane for radio (wireless) instruction.
Armament: One VK II canister on the upper wing.
65.80: On 9 March 1917 made available to *Feldpilot Hptm.* Josef Cejnek at *Übernahmsgruppe* I (Acceptance Group I) in Aspern for trial purposes; later a pursuit trainer with *Flek* 6. Still on the rolls of the *Flet* as of 31 August 1918.
65.81: Beginning of February 1917 to *Fluggeschwader* I, traceable until May 1917; it was damaged in September 1917 following a landing accident; written off October 1917.

KD pilots, Petzold seated in center, *Hptm.* Raoul Stojsavljevic standing at far left, others unknown. Photo: Walter Schroeder.

Above: Close-up of the engine of a KD series 65.70. Photo: Walter Schroeder.
Left: KD series 65.70. Photo: Walter Schroeder.

Left: KD 65.56 on right and 65.74 on left. Photo: Walter Schroeder.

Below left: KD 65.75. Photo: Walter Schroeder.

65.82: Delivered 1917, written off March 1918.
65.83: Delivered 1917 as a pursuit trainer with *Flek* 6; still found on the rolls of the *Flet* as of 31 August 1918.
65.84: Delivered 1917 as a pursuit trainer with *Flek* 6, still found on the rolls of the *Flet* as of 31 August 1918.
65.85: Delivered 1917 as a pursuit trainer with *Flek* 6; crashed during a training flight and was written off 10 May 1917.
65.86: Delivered 1917 as a pursuit trainer with *Flek* 6, still found on the rolls of the *Flet* as of 31 August 1918.
65.87: Delivered 1917 as a pursuit trainer with *Flek* 6, written off August 1918, following an accident due to fog.
65.88: Delivered 1917 as a pursuit trainer with *Flek* 6. On 3 November 1917, *Lt.* Leder crashed fatally during a practice flight, flying into the ground in a steep dive with full power shortly before landing at Theresienfeld. The aircraft burned and was written off in January 1918. Despite being written off, on the rolls of the *Flet* as of 31 August 1918.
65.89: Delivered 1917 as a pursuit trainer with *Flek* 6, still found on the rolls of the *Flet* as of 31 August 1918.
65.90: To *Flik* 16 on 28 April 1917, determined in fact to be at *Flik* 41/J in May 1917. Sent to the rear by the *Flep* on 6 October 1917.
65.91: To *Flik* 41/J as a

The crash of *Feldpilot Oberleutnant* Gustav Rubritius of *Flik* 34 with KD 65.78 on 31 January 1917. See page 60 for another view of this crash. Photo: War Archive, Vienna.

Zugsführer Karl Seiler of *Flik* 35 (and mascot) in his KD 65.79 at St.Veit (near Wippach) in 1917. See page 60 for another view. Photo: Walter Schroeder.

pursuit trainer on 28 April 1917; from July 1917 with *Flik* 4 for training purposes.
65.92: Delivered 1917 as a pursuit trainer with *Flek* 6. Written off 20 July 1918, it was nevertheless listed on the rolls of the *Flet* on 31 August 1918. In 1919, it was determined as still being in Austria.
65.93: Delivered 1917 as a pursuit trainer with *Flek* 6; mentioned as being on the rolls of the *Flet* as of 31 August 1918.
65.94: Delivered 1917 as a pursuit trainer with *Flek* 6; written off after a landing accident in September 1917 (see page 12).
65.95: Delivered 1917 as a pursuit trainer with *Flek* 6; written off January 1918 after a landing accident. Nevertheless, it is mentioned as being on the rolls of the *Flet* on 31 August 1918.
65.96: Delivered 1917 as a pursuit trainer with *Flek* 6; still found on the rolls of the *Flet* as of 31 August 1918.
65.97: Delivered 1917, determined as being a pursuit trainer with *Fleks* 6 and 5; mentioned as still being on the rolls of the *Flet* as of 31 August 1918.
65.98: Delivered 1917 as a pursuit trainer with *Flek* 6; written off March 1918; it was nevertheless mentioned in the rolls of the *Flet* as of 31 August 1918.
65.99: Delivered 1917; in June 1918, it was listed as a pursuit trainer without an engine with *Flek* 6.

Below: NCO pilots of *Flik* 23 with KD 65.67, Prosecco airfield, Summer 1917. Photo: Karl Meindl.

Left: KD 65.67.

Below: KD series 65.7.

Left: Engine installation in a KD series 65.7.
Below: KD series 65.7.

Above and above left: KD 65.71; used as a fighter trainer with *Flek* 6.

KD 65.72; used as a fighter trainer with *Flek* 6.

Another view of KD 65.72.

Four above photos of KD 65.77, which was used as a fighter trainer at *Flek* 6. The rudder and fin of KD 65.77 have been modified to improve stability.

Above: The crash of KD 65.78 at *Flik* 34. See page 56 for another view of this crash.

Above left: Another view of KD 65.72 at *Flek* 6.

Left: Hansa-Brandenburg 65.79 with pilot *Zugsführer* Karl Seiler of *Flik* 35. See page 57 for another view of this aircraft.

The War Diary of *Feldpilot* *Oberleutnant* Frank Linke-Crawford

(When flying the KD)

January 1917: Flek 6, Wiener-Neustadt, Pilot Training

[Probably with Series 65.5 and 65.7]
13 Jan.: 9 training flights with a KD
14 Jan.: 4 training flights with a KD
15 Jan.: 1 training flight with a KD
22 Jan.: 2 training flights with a KD
January 1917: *Flik* 12, Wippach Airfield, Isonzo Front
26 Jan.: Test flight with 65.53, heavily damaged during landing due to ground wind gusts (Undercarriage and 2 wing ribs broken), pilot uninjured.
31 Jan.: 2 training flights with a KD

February 1917:

4 Feb.: Training flight with a KD (65.51 or 65.59, airplanes of *Flik* 19)
Training flight with a Fokker single-seater (both flights with *Flik* 19)
10 Feb.: Combat patrol with a KD (Number of the KD aircraft flown is unknown)
11 Feb.: 2 combat patrols with a KD (Number of the KD aircraft flown is unknown)
13 Feb.: 2 combat patrols with a KD (Number of the KD aircraft flown is unknown)
14 Feb.: Combat patrol with a KD (Number of the KD aircraft flown is unknown)
16 Feb.: Combat patrol with a KD (Number of the KD aircraft flown is unknown)
Combats with enemy airplanes
24 Feb.: Test flight with KD 28.02
25 Feb.: 3 combat patrols with KD 28.02, one combat with an enemy airplane
26 Feb.: Combat patrol with KD 28.02
27 Feb.: Test flight with KD 65.55 of *Flik* 34
Combat patrol with KD 28.02, turned back because of heavy clouds
28 Feb.: Combat patrol with KD 28.02

March 1917:

1 Mar.: Combat patrol with KD 28.02
10 Mar.: 2 combat patrols with KD 28.02
11 Mar.: Combat patrol with KD 28.02
17 Mar.: Combat patrol with KD 28.02, combat with a Caudron G.IV
18 Mar.: Combat patrol with KD 28.02, combat with 2 Nieuport single-seaters
Combat patrol with KD 28.02, combat with 2 Caudron G.IVs and 4 Nieuport single-seaters
19 Mar.: Combat patrol with KD 28.02

April 1917:

5 Apr.: Combat patrol with KD 28.02
12 Apr.: Combat patrol with KD 28.02
13 Apr.: Combat patrol with KD 28.02, combat with a Caudron, later a combat with 2 Nieuport single-seaters
17 Apr.: Combat patrol with KD 28.02, turned back because of heavy clouds
19 Apr.: Combat patrol with KD 28.02, during which face was frostbitten due to the intense cold, sick until 24 April.
25 Apr.: Combat patrol with KD 28.02

May 1917:

1 May: Combat patrol with KD 28.02, 3 combats, with a Caudron G.IV and 2 Nieuport single-seaters
Combat patrol with KD 28.04 of *Flik* 34 (with new tail surfaces)
3 May: Combat patrol with KD 28.02, 2 combats with a Nieuport single-seater and a Caproni bomber
Combat patrol with KD 28.02, a combat with a Nieuport single-seater
5 May: Combat patrol with KD 28.02, combats with Farman and Nieuport escort planes, own machine gun jammed
Combat patrol with KD 28.02
10 May: Escort flight, combat over Cormons with a Farman, own machine gun jammed
11 May: Forced to land, Wippach airfield with KD 28.02
Defensive patrol with KD 28.02, machine gun jammed
Combat patrol with KD 28.02
12 May: Combat patrol with KD 28.02, 9 enemy aircraft pursued, combat with 1 Caudron, 2 Nieuports, and 1 SPAD. Machine-gun hit to 28.02, 1 round through the leather of the pilot's seat
16 May: KD 28.02 test-flown with new stabilizers.
Line patrol with KD 28.02, 2 enemy airplanes driven back
18 May: Combat patrol with KD 28.02, combat with a biplane
Combat patrol with KD 28.02
Line patrol with KD 28.02
19 May: Line patrol with KD 28.02
21 May: Escort flight for reconnaissance planes of *Flik* 4, combat with a Caudron G. IV and a Nieuport
25 May: Line patrol with KD 28.02
Line patrol with KD 28.02, combat with a SPAD, 14 hits to 28.02

Oblt. Frank Linke-Crawford; a pencil drawing by Theo Matejko made at Feltre airfield, June 1918. The signature at the bottom reads "Frank Linke-Crawford, *Oblt.*" With 27 victories Linke-Crawford was the fourth ranking Austro-Hungarian ace. His mother was English.

Linke-Crawford with KD 28.15, *Flik* 41/J, Sesana, 4 August 1917. There is a small serial number on the wheel cover near the hub. Linke-Crawford was KIA on 31 July 1918, apparently after structural failure of his aircraft, Aviatik D.I 115.32, rendered him unable to take evasive action.

June 1917:

8 June: New KD 28.38 test flown
17 June: Escort flight with KD 28.38, bombing raid on Mossa, combat with a Nieuport fighter
18 June: Line patrol with KD 28.38, combat with a Nieuport fighter
19 June: Escort flight for photo reconnaissance aircraft of *Flik* 12, with KD 28.38
Defensive patrol with KD 28.38
20 June: Line patrol with KD 28.38
21 June: Line patrol with KD 28.38, together with *Feldpilot Feldwebel* Alexander Renner of Flik 34 in KD 28.04, combat with Nieuport, own machine gun jammed
22 June: Line patrol with KD 28.38
25 June: Line patrol with KD 28.38, together with *Feldwebel* A. Renner (in KD 28.04) 3 combats with 4 Nieuports
26 June: Line patrol with KD 28.38, together with *Feldwebel* Karl Seiler of *Flik* 35 (in KD 28.43)
Line patrol with KD 28.38
28 June: Escort flight for photo reconnaissance aircraft of *Flik* 12, with KD 28.38
Line patrol with KD 28.38, combats with 2 Caudrons and Nieuports
30 June: Line patrol with KD 28.38, own machine gun jammed, turned back

July 1917:

13 July: Combat patrol with KD 28.38, combat with a "lattice-tail"
14 July: Combat patrol with KD 28.38, combat with a "lattice-tail"
17 July: Defensive patrol with KD 28.38
19 July: Combat patrol with KD 28.38, together with *Feldwebel* Renner of *Flik* 34 (in KD 28.04)
20 July: Escort flight with KD 28.38, bombing raid on Cividale
Defensive patrol with KD 28.38, combat with enemy airplane
24 July: Escort flight with KD 28.38
26 July: Escort flight with KD 28.38
28 July: Escort flight with KD 28.38, combat with a SPAD fighter
30 July: Escort flight with KD 28.38
New KD 28.66 test flown

August 1917:

1 Aug.: New KD 28.66 test flown
4 Aug.: Linke-Crawford to *Flik* 41/J as fighter pilot

August 1917: *Flik* 41/J

4 Aug.: During a test flight, KD 28.15 damaged upon landing on the Sesana airfield (see photo above)
5 Aug.: Escort flight with KD 28.44, fired on from long range by a SPAD
6–16 Aug.: in Wiener-Neustadt for additional training with *Flek* 6
21 Aug.: Combat patrol with KD 28.40, combats with a Caproni and Nieuport, first air victory
Combat patrol with KD 28.13, combats with a "lattice-tail" and a Nieuport over Kostanjevica, later a combat with a "lattice-tail" over Monte Santo—Plava
22 Aug.: Combat patrol with KD 28.40, combat with a "lattice-tail", second air victory
Combat patrol with KD 28.40, several combats with enemy aircraft
23 Aug.: Combat patrol with KD 28.40, shot down a Savoia-Pomilio, third air victory
Combat patrol with KD 28.40, combats with Caproni and SPAD aircraft
24 Aug.: Combat patrol with KD 28.40, combats with SPADs
Combat patrol with KD 28.40, combats with a SPAD, a Nieuport, and a Caudron
25 Aug.: Combat patrol with KD 28.40, combat with a "lattice-tail", forced to land
Combat patrol with KD 28.40, combat with a Sopwith two-seater
26 Aug.: Combat patrol with KD 28.40, 5 combats, with 4 SPADs and a "lattice-tail", shot down a SPAD fighter, fourth air victory
29 Aug.: Combat patrol with KD 28.40, combat with a Nieuport
30 Aug.: Escort flight with KD 28.40, emergency landing due to an engine defect
31 Aug.: 2 combat patrols with KD 28.40

September 1917:

1 Sep.: Combat patrol with KD 28.40, combats with enemy aircraft
6 Sep.: Combat patrol with KD 28.40, combat with a SPAD
7 Sep.: 3 combat patrols with KD 28.40

From 8 September 1917, Linke-Crawford flew other fighter types such as the Albatros D.III, Phönix D.I, and Aviatik D.I.

Acknowledgements

I would like to thank the following people for their reliable and kind assistance:

Mr. Peter M. Grosz, U.S.A
Dr. Roberto Gentilli, Firenze, Italy
Mr. Johannes Heubel, Ansfelden, Austria
Mr. Oswald *Freiherr* Kostrba-Skalicky, Vienna, Austria
Mr. Gino Piccoli, Rovereto, Italy
Hofrat (Counsellor) Dr. Peter Schiemer, Vienna, Austria
Mrs. Herta Schroeder, Vienna, Austria
Mr. Jörg C. Steiner, Vienna, Austria
ObIng. Franz Selinger, Ulm, Germany

Index of Sources

a) The Austrian State Archives/War Archives, Vienna, Austria:

Fascicle 19–22 (Orders of the Day of the *k.u.k. Luftfahrtruppen* 1916/1917)
Fascicle 70–111 (Pilots' personal files and combat flight logs)
Fascicle 165 (Personal file and combat flight log of Fiala-Fernbrugg)
Fascicle 165 (Personal file and combat flight log of *FP* Arigi)
Fascicle 166 (Personal file and combat flight log of Brumowski)
Fascicle 167 (Personal file, combat flight log and diary of *FP* Frank Linke-Crawford)
Fascicle 181 (Reports from front-line positions 1917/1918)
Fascicle 182–188 (Monthly reports of the *k.u.k. Fliegerkompagnien*)
Fascicle 290–291 (Ten-day reports of the *Fliegerkompagnien* 1916)
Fascicle 387 (Air combats of the Austro-Hungarian *Fliegerkompagnien*)
Fascicle 388 (Air victories of the *k.u.k. Fliegerkompagnien*)
Nachrichtenblatt der k.u.k. Luftfahrtruppen 1917/1918
Army and Navy reports 1916/1917
Numerous award requests of KD pilots
Various individual files

b) Private Sources in Austria and Abroad

Letter correspondence of former *k.u.k.* airmen (K. Meindl Archive)
Archive of *ObIng.* Franz Selinger, Ulm, Germany
Archive of Rinaldo d'Ami, Mailand, Italy
Archive of Gino Piccoli, Rovereto, Italy
Archive of Dr. Roberto Gentilli, Firenze, Italy
Archive of Walter Schroeder, Vienna, Austria
Archive of Jörg E. Steiner, Vienna, Austria
Archive of Bernhard Tötschinger, Vienna, Austria

c) Unpublished Studies

KD-Flugzeugliste by *Hofrat* Dr. Peter Schiemer
Beute-Flugzeuge by *Hofrat* Dr. Peter Schiemer
Die Luftsiege der k.u.k. Fliegertruppe 1915–1918 an der Italienischen Front by Karl Meindl
Die Luftsiege der k.u.k. Fliegertruppe 1916–1917 an der Rumänischen Front by Karl Meindl

d) Literature

Die Flugzeuge der k.u.k. Luftfahrtruppe und Seeflieger 1914–18, E. Hauke, W. Schroeder, B. Töschinger, H. Weishaupt Verlag, Graz, 1988.

Austro-Hungarian Army Aircraft of World War One, P. Grosz, G. Haddow, Dr. P. Schiemer, Flying Machines Press, USA, 1993.

Glossary and Abbreviations

ausser Dienst (a.D.): Retired.
Bauart (BA): Construction design.
Beobachteroffizier (BO): Observation officer.
Bora: A cold wind storm, sometimes of great destructive force, especially on the northern shores of the Adriatic.
Divisionsfliegerkompagnie (D-Flik): A short-range reconnaissance unit working in conjunction with army divisions at the front.
Doppeldecker: Biplane.
Feldpilot (FP): Field Pilot, a title awarded after approximately ten combat flights.
Fernaufklärungskompagnie (F-Flik): Long-range reconnaissance unit.
Flieger Etappenpark (Flep): Rear-area aviation park. A major rear-area aircraft supply and repair park. Five were formed in August 1914, later increased to eleven. Renamed *Fliegerpark (Flip)* in June 1918.
Fliegerarsenal (Flars): Aviation arsenal. The organization responsible for testing, production, and acceptance of all army aviation and balloon equipment; aircraft and material supply; technical development of aircraft, engines, bombs, armament, cameras, propellers, etc.
Fliegerersatzkompagnie (Flek): An aviation replacement company. Training unit for primary and advanced flight instruction, for training observers, mechanics, and other aviation personnel. Based throughout the Dual Monarchy, at war's end there were 22 active *Fleks*. Some *Fleks* were responsible for organizing and outfitting newly-formed *Fliks* prior to the *Fliks* leaving for the front.
Fliegerersatztruppen (Flet): Aviation replacement troops. Basically the aviation training command.
Fliegerkompagnie (Flik): An aviation company. The *Flik* was the basic operational or fighting unit of the *Luftfahrtruppe*. In July 1917, *Fliks* were assigned specialized combat tasks and identified accordingly.
Fliegertruppen (Flet): Aviation troops.
Fliegerpark (Flip): See *Fliegeretappenpark*.
Fluggeschwader (FlG): Flying squadron.
Fluggeschwader I (FlG I): Bombing unit, from September 1917 designated 101/G.
Flugwache: Observation post.
Frieherr: Baron
in der Reserve (i.d.Res.): In the reserves.
Jagdfliegerkompagnie (J-Flik): Fighter aviation detachment or unit.
Kaiserlich und königliche (k.u.k.): Imperial and Royal.
Kaiserlich-königlich (k.k.): Imperial-Royal (used by the Austrian *Landwehr*).
Kommandant der Luftfahrtruppen (Koluft): Officer in charge of the aviation troops assigned to an army.
Luftfahrtruppe (LFT): Aviation troops.
Ritter: Knight, equivalent to the British title "Sir", shows elevation to the nobility.
Schwarzlose: The standard machine gun of the Austro-Hungarian Army, a lighter-weight, air-cooled version of which was developed for use by the Austrian Air Service.
Spandau: The common term used to denote the German-made Maxim machine gun; it derives from the name of the Berlin suburb where the factory was located.
Stabsoffizier der Luftfahrtruppen (Stoluft): Staff officer of the aviation troops assigned to an army.
Übernahmesgruppe: Acceptance Group.
VK II: Machine-gun canister.

Austro-Hungarian Ranks

Feldmarschalleutnant: Field Marshal Lieutenant
Feldwebel (Feldw.): Sergeant or sergeant major, lowest noncommissioned officer rank.
Generaloberst: Colonel General.
Hauptmann (Hptm.): Captain.
Korporal (Korp.): Corporal, an enlisted rank.
Leutnant (Lt.): Second Lieutenant (U.S.).
Linienschiffsleutnant: Lieutenant Commander.
Oberleutnant (Oblt.): First Lieutenant (U.S.).
Oberst: Colonel.
Offizierstellvetreter (Offstv.): Acting officer, highest noncommissioned officer rank.
Rittmeister (Rittm.): A cavalry Captain (literally "riding master").
Stabsfeldwebel (Stabsfeldw.): Staff sergeant, noncommissioned officer rank.
Zugsführer (Zugsf.): Platoon leader or sergeant, an enlisted rank.

Aircraft

Caproni: A tri-motor Italian bomber built in different versions.
Caudron: A French reconnaissance type built in several versions, two of which were used by the Italians.
Farman: A French reconnaissance type of pusher configuration built in several versions, some used by the Italians.
Hansa-Brandenburg C.I: The most-widely used Austro-Hungarian reconnaissance aircraft type.
Kampfdoppeldecker (KD): A biplane fighter plane, usually meaning the Hansa-Brandenburg D.I.
Nieuport: A French-designed single-seat fighter used by the Italians. Some were built under license in Italy.
SAML: Italian reconnaissance type.
Savoia-Pomilio (S.P.): An Italian reconnaissance type built in several versions. The abbreviation S.P. (according to the production company and designer) was used by the Italian airmen with the following meanings, due to its high loss rate: *Sepoltura **p**er due* = a coffin for two
*Siamo **p**erduti* = we are lost, done for
SPAD: A French-designed single-seat fighter used by the Italians. Some were built under license in Italy.
Voisin: A French reconnaissance type of pusher configuration built in several versions, some used by the Italians.

Italian Military Ranks 1915–1918

Aspirante ufficiale = Officer Candidate
Capitano = Captain
Caporale = Corporal
Caporale maggiore = Sergeant, platoon leader
Colonello = Colonel
Maggiore = Major
Maresciallo = Acting Officer
Mitragliere = Machine Gunner
Osservatore = Observer Officer
Pilota = pilot
Sergente = Sergeant
Sergente maggiore = Staff Sergeant
Soldato = Soldier
Sottotenente = Second Lieutenant
Tenente = First Lieutenant
Tenente colonello = Lieutenant Colonel

Italian Naval Ranks 1915–1918

Enlisted Men / Noncommissioned Officers

Marinaio = Seaman
Sottocapo = Petty Officer
Capo = Chief Petty Officer

Officers

Guardiamarina = Midshipman
Sottotenente di vascello = Naval Second Lieutenant
Tenente di vascello = Naval Lieutenant

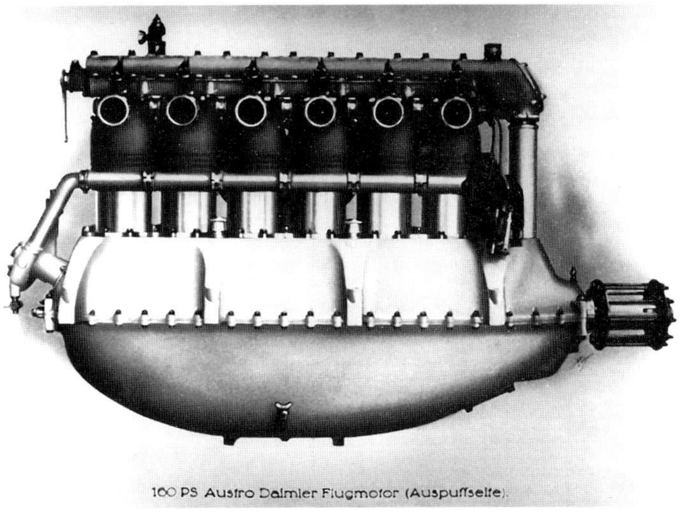

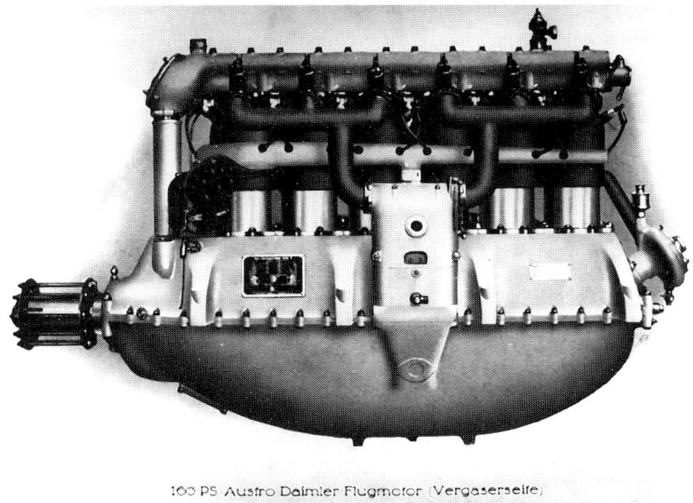

Above and above right: Two views of the 160 hp Austro-Daimler engine. Photos: Josef Bangerl via Karl Meindl.

Color profile references and captions: Walter Schroeder
Translation: Jan Hayzlett
Color profiles: Juanita Franzi
Cover painting: Joe Keogan
Scale drawings: Martin Digmayer
Book and cover design and layout: Jack Herris
Digital scanning and editing: Gretchen Herris
Photos: Unless otherwise credited, Peter M. Grosz
ISBN: 1-891268-01-5

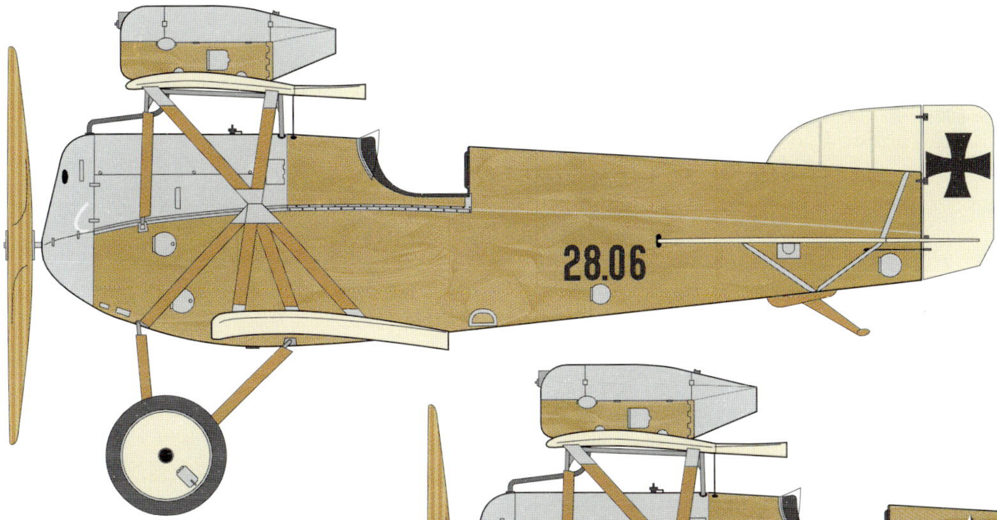

28.06, *Fluggeschwader* I. Three confirmed air victories by *FP Offstv.* Julius Arigi. Also flown by pilots Bakulinsky, Dovolil, and Strohschneider.

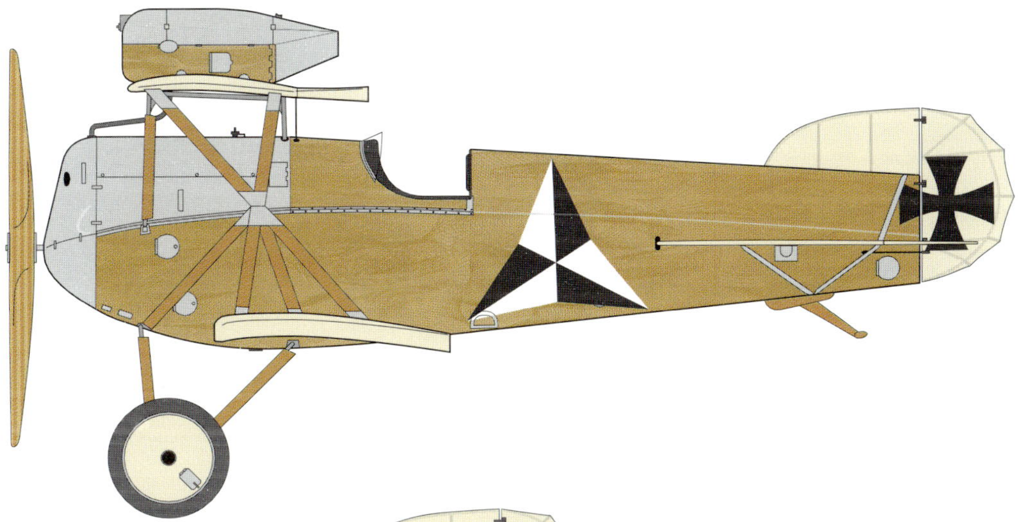

28.10, *Flik* 41/J, flown by Brumowski, Samek, Richter, Mayrbäurl, and Pillwein, among others, shown here after retrofit with new fin and rudder.

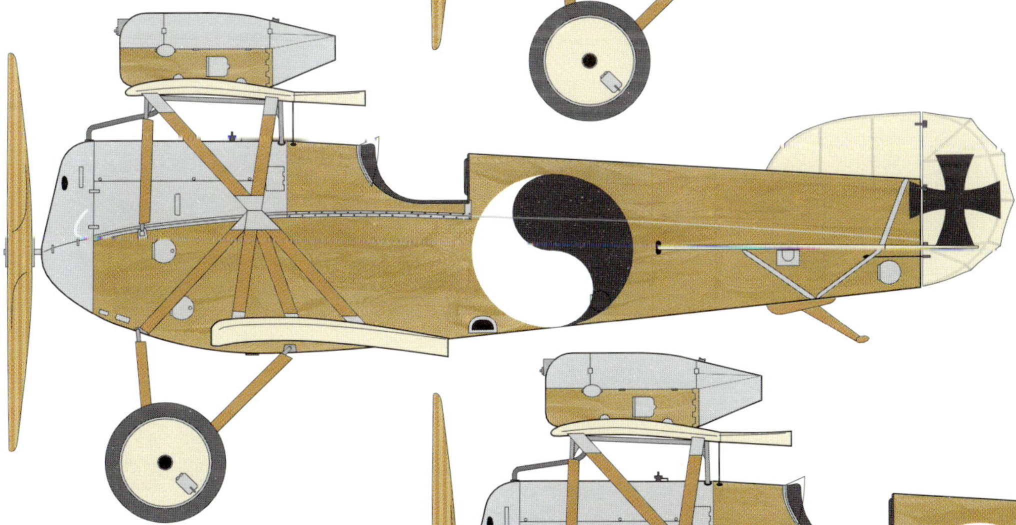

28.11, *Flik* 41/J, written off in July 1917 by *Feldwebel* Karl Kaszala.

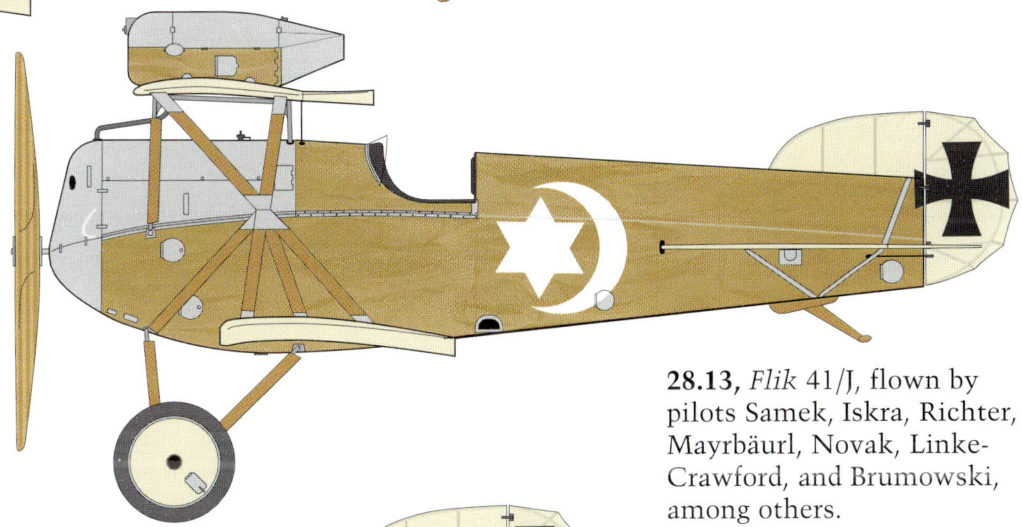

28.13, *Flik* 41/J, flown by pilots Samek, Iskra, Richter, Mayrbäurl, Novak, Linke-Crawford, and Brumowski, among others.

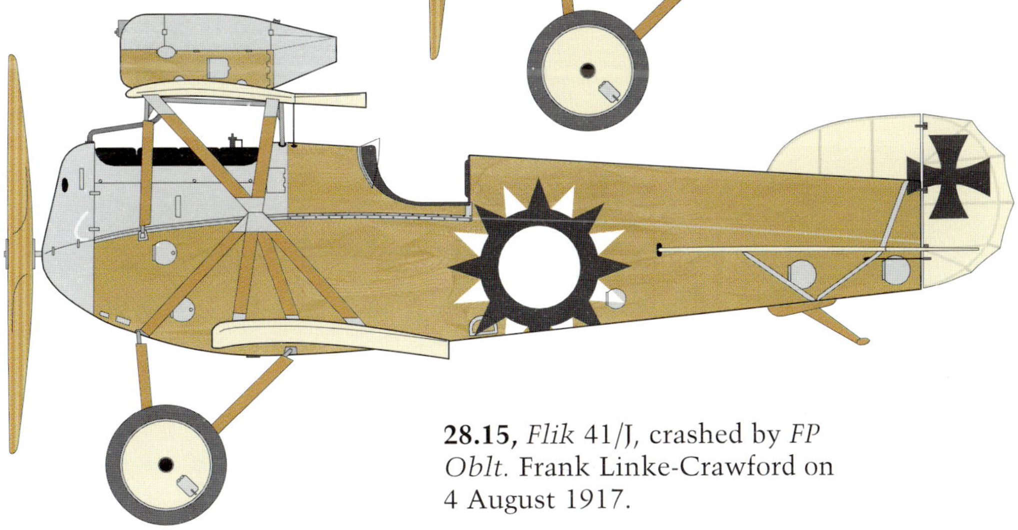

28.15, *Flik* 41/J, crashed by *FP Oblt.* Frank Linke-Crawford on 4 August 1917.

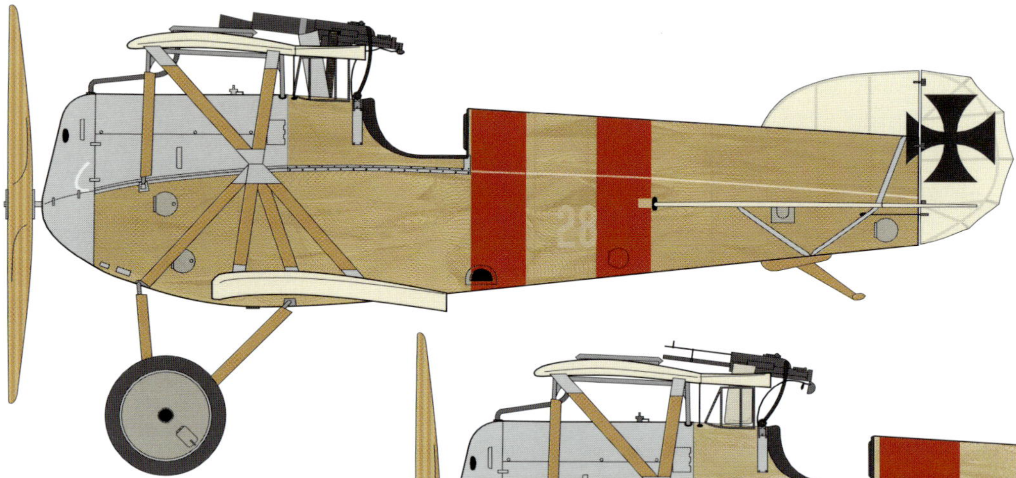

28.26, *Flik* 16, flown by pilots *Hptm.* Stojsavljevic, *Oblt.* Urban, *Offstv.* Siegel, and *Stfw.* Ehret. On 12 January 1918, *Hptm.* Stojsavljevic was shot down in air combat and was both injured and badly wounded.

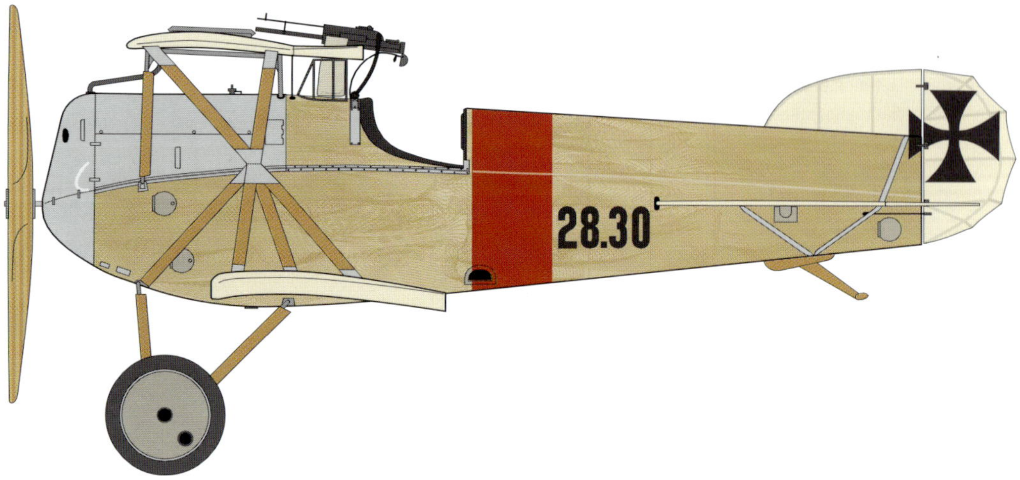

28.30, *Flik* 16, an angled MG 7/12. On 25 May 1917 *FP Fw.* Erwin Funk achieved an air victory. In addition, *Hptm.* Stojsavljevic achieved two air victories with this airplane.

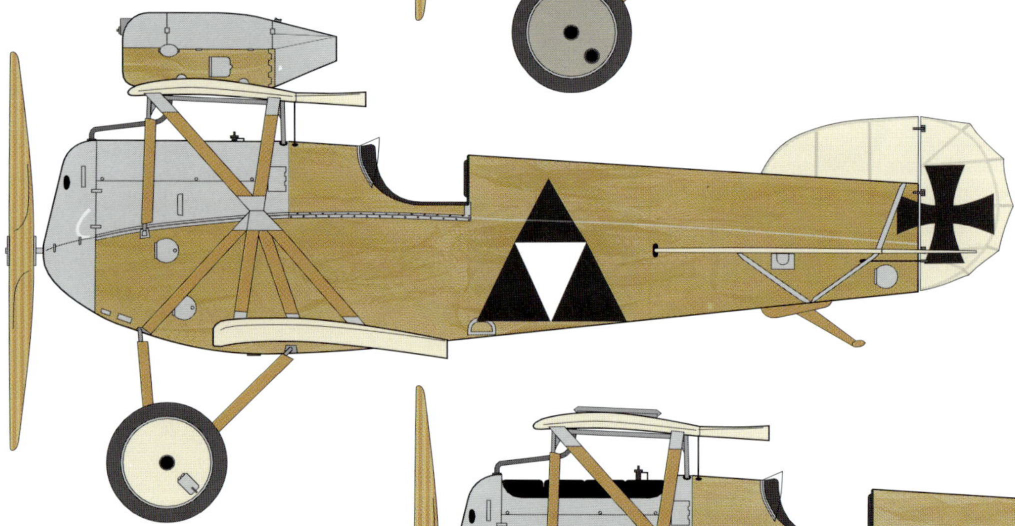

28.33, *Flik* 41/J, flown by pilots *Hptm.* Brumowski, *Oblt.* Fiala, and *Offzstv.* Arigi. One air victory by *FP Zugsf.* Josef Novak, a second air victory on 17 July1917.

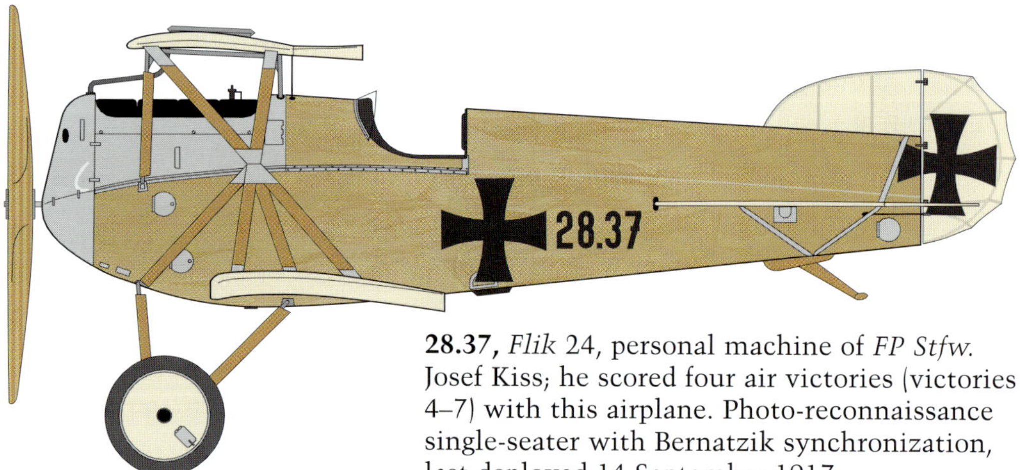

28.37, *Flik* 24, personal machine of *FP Stfw.* Josef Kiss; he scored four air victories (victories 4–7) with this airplane. Photo-reconnaissance single-seater with Bernatzik synchronization, last deployed 14 September 1917.

28.38, *Flik* 12, one unconfirmed air victory by *FP Oblt.* Frank Linke-Crawford; four confirmed and several unconfirmed air victories by *FP Oblt.i.d.Res.* Benno Fiala *Ritter* von Fernbrugg.

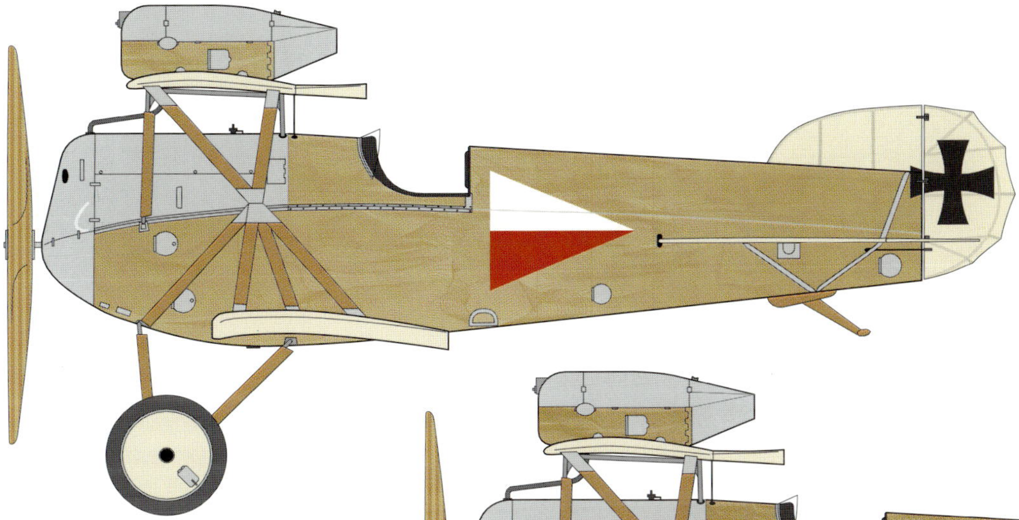

28.40 *Flik* 41/J, personal machine of *FP Oblt.* Frank Linke-Crawford (four confirmed and one unconfirmed air victories). Flown by pilots Fischer, Fiala, Brumowski, Samek, Novak, Mayrbäurl, and Pillwein. In addition, *FP Fw.* Hermann Richter scored an air victory in this machine on 20 August 1917.

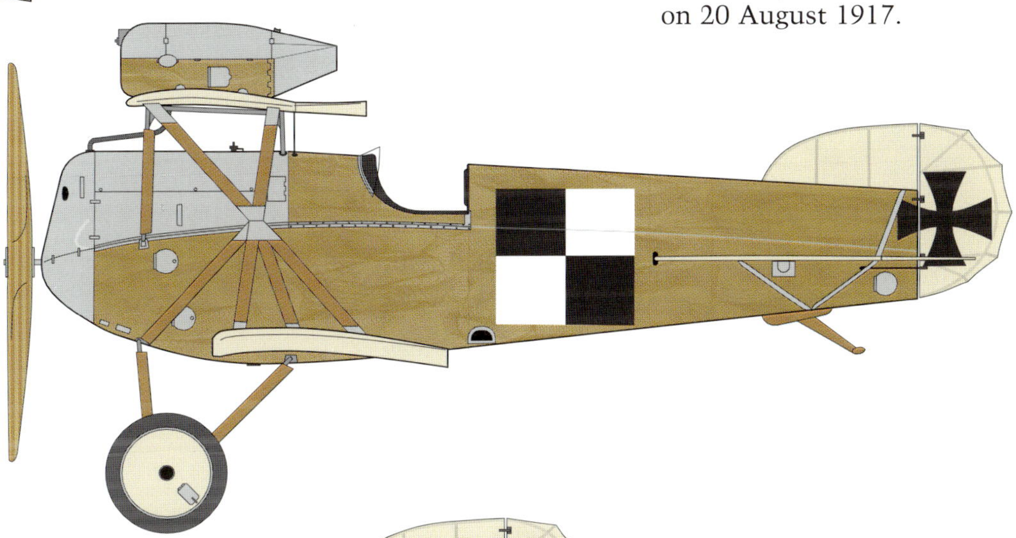

Flik 41/J, checkerboard pattern, presumably in black and white.

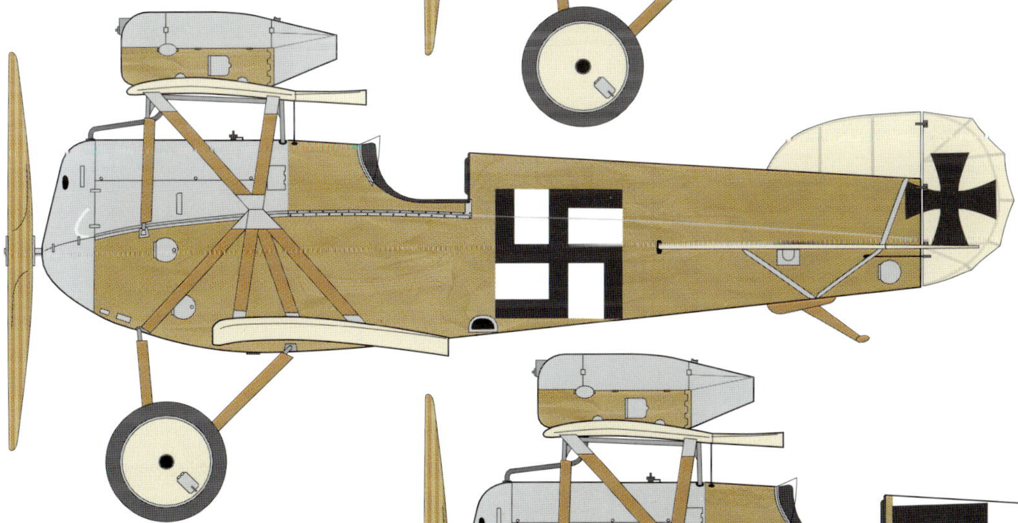

Flik 41/J, aircraft number unknown, with the protective religious symbol of the Buddhists—"Come home safe".

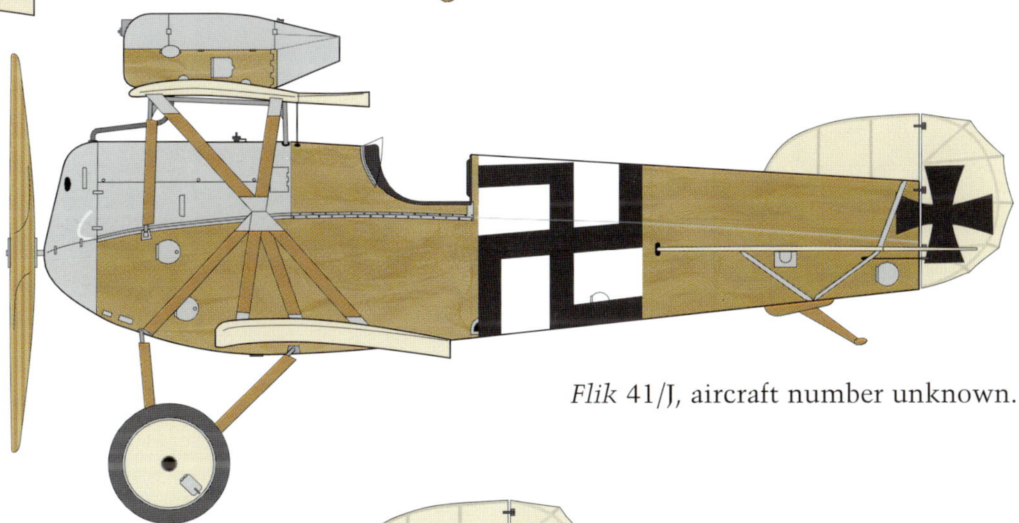

Flik 41/J, aircraft number unknown.

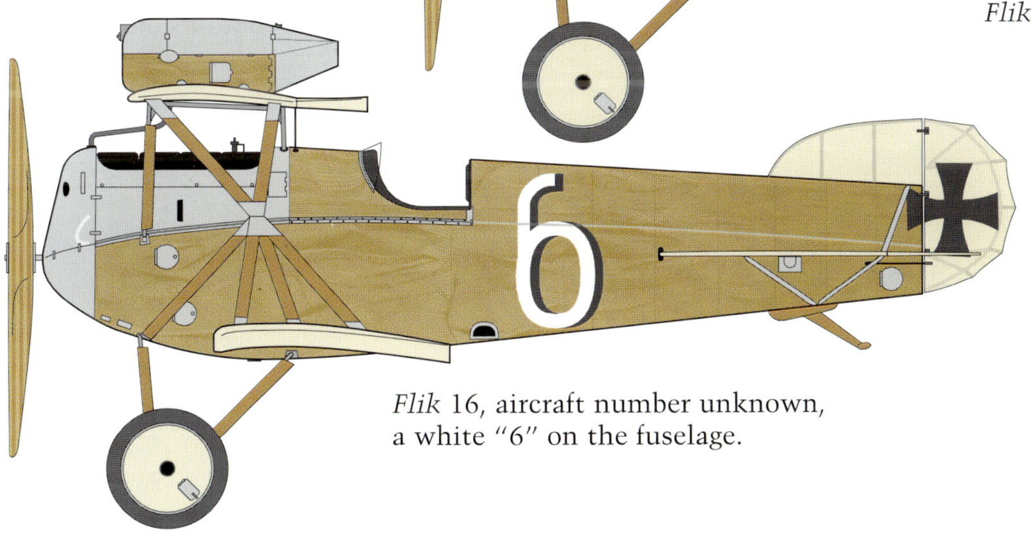

Flik 16, aircraft number unknown, a white "6" on the fuselage.

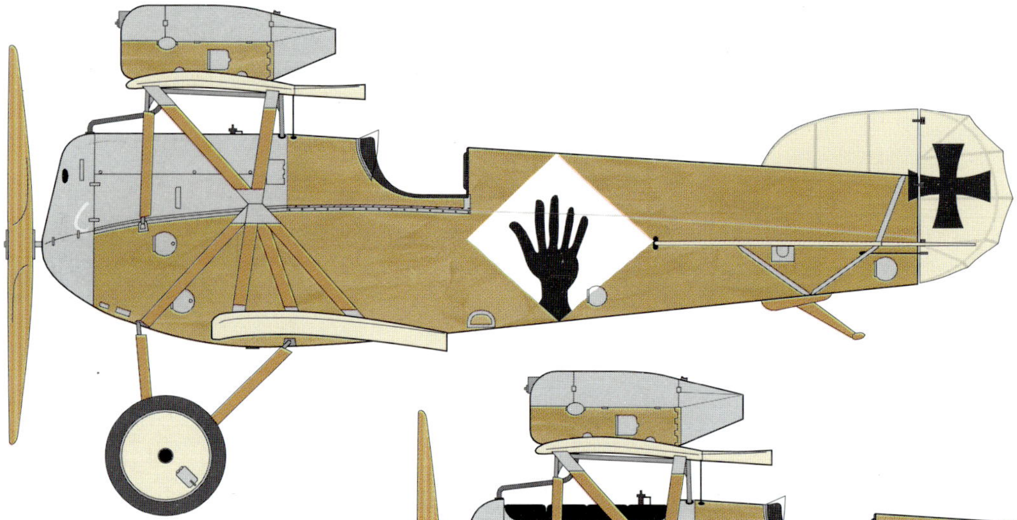

Flik 41/J, **28.44** (?). *Korp.* Altmann Herzog crashed to his death on 16 December 1917 as the result of sabotage.

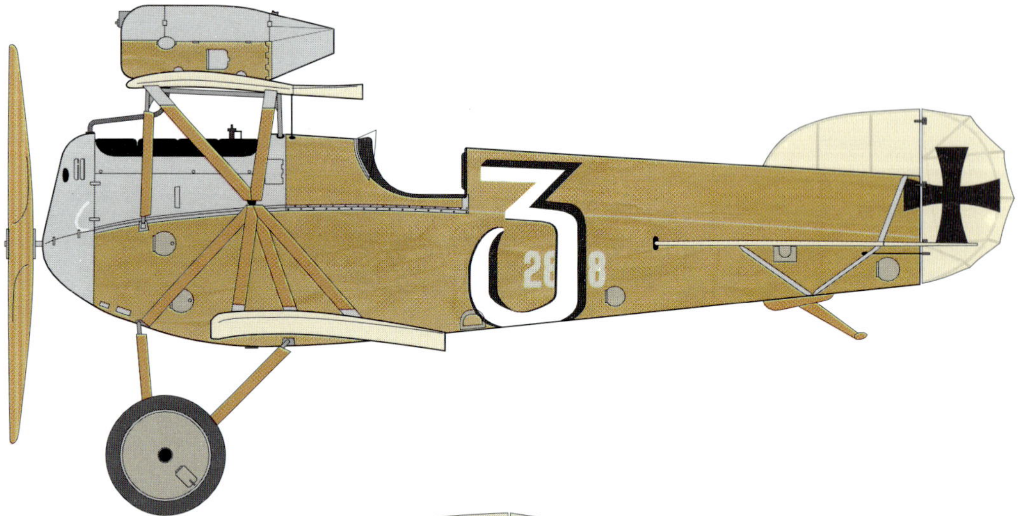

KD 28.58, *Flik* 16, a white "3" painted on the fuselage. Airplane was with *Flik* 42/J in July 1917, was then damaged, and then went to *Flik* 16 in Kärnten by way of *Flep* 1 in Marburg.

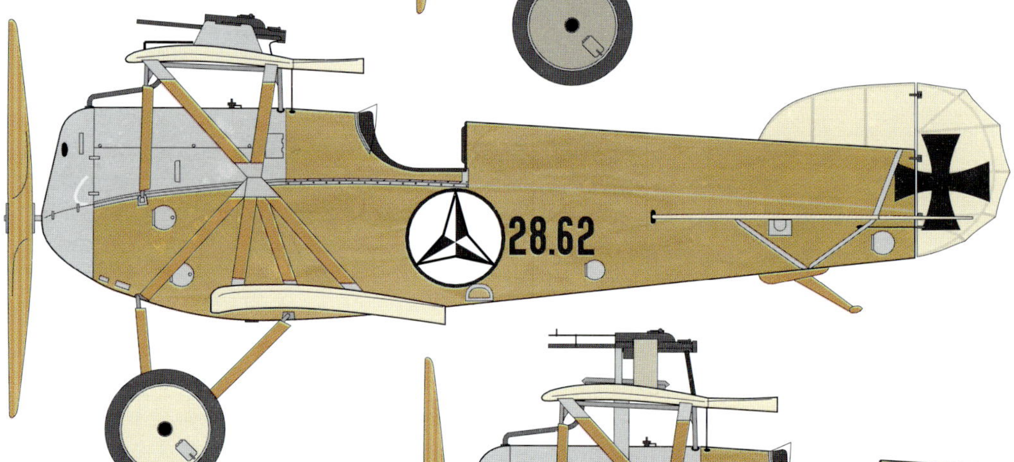

28.62, *Flik* 46, personal machine of *Oblt.i.d.Res.* Dr. Josef Hoffmann *Ritter* von Ostenhof.

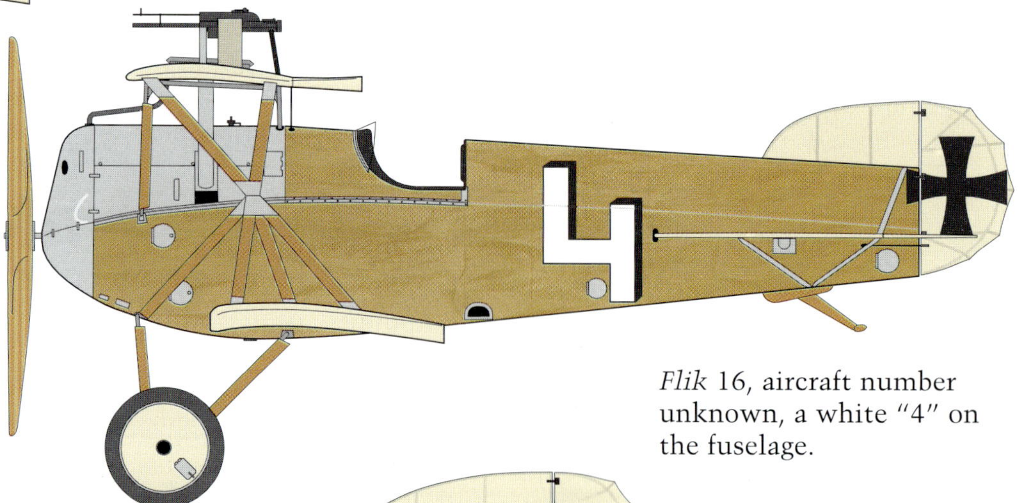

Flik 16, aircraft number unknown, a white "4" on the fuselage.

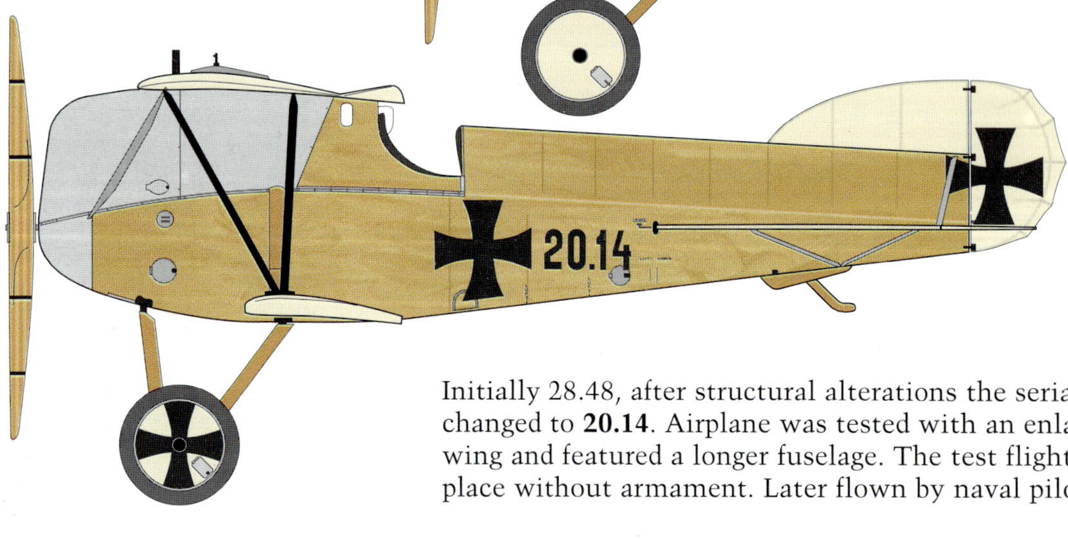

Initially 28.48, after structural alterations the serial was changed to **20.14**. Airplane was tested with an enlarged wing and featured a longer fuselage. The test flights took place without armament. Later flown by naval pilots.